I0492113

The Home Poultry Book

by Edward I. Farrington

with an introduction by Jackson Chambers

This work contains material that was originally published in 1913.

This publication is within the Public Domain.

*This edition is reprinted for educational purposes
and in accordance with all applicable Federal Laws.*

Introduction Copyright 2018 by Jackson Chambers

THE WORLD'S LARGEST SELECTION OF VINTAGE POULTRY BOOKS

www.VintagePoultry.com

Introduction

I am pleased to present yet another title on Poultry.

The work is in the Public Domain and is re-printed here in accordance with Federal Laws.

As with all reprinted books of this age that are intended to perfectly reproduce the original edition, considerable pains and effort had to be undertaken to correct fading and sometimes outright damage to existing proofs of this title. At times, this task is quite monumental, requiring an almost total "rebuilding" of some pages from digital proofs of multiple copies. Despite this, imperfections still sometimes exist in the final proof and may detract from the visual appearance of the text.

I hope you enjoy reading this book as much as I enjoyed making it available to readers again.

Jackson Chambers

Self Reliance Books

Get more historic titles on animal and stock breeding, gardening and old fashioned skills by visiting us at:

http://selfreliancebooks.blogspot.com/

PREFACE

NO doubt the experts whose eyes chance to fall on this book will say that it is elementary. It is, and purposely so. It is designed first and last for the amateur who has no time or inclination to read technical and semi-technical books on poultry keeping. It aims to tell the man with a few hens what to do and how to do it. There are certain statements with which some poultry keepers will disagree. We may look for the millennium when the men who keep hens come to be of one mind as to the management of them. However, no pet theory has been put forward or new system advocated. What has been written is based largely on personal experience and what has been learned in visits to many poultry plants, large and small, in different parts of the country. It is sent out with the hope that some amateurs may find in it a suggestion or two which will help them to get more eggs with less expense and have more fun doing it.

CONTENTS

THE ILLUSTRATIONS

THE ILLUSTRATIONS

THE HOME POULTRY BOOK

Chapter I

HOW TO MAKE A BEGINNING

WHEN a man — or for that matter, a woman — is smitten with the poultry fever, he usually knows little about breeds or methods or "systems." Only one fact presents itself — he wants to keep a few hens.

Too often the beginner makes so little distinction between hens in general and those of particular breeds, that he accumulates a flock of mongrels. Beginning in that way, he may, perhaps, get as many eggs in a year as though he had started with a flock of pure-bred fowls, but the chances are that he will soon tire of poultry keeping. It is impossible to work up much enthusiasm over a lot of birds which have uniformity neither in color nor size and which show in their feathers an intermixture of many breeds and varieties.

I

On the other hand, there is rare pleasure in watching the movements of a little flock in which every cock and hen has the size, shape and markings typical of a recognized breed. For that reason, and because his poultry keeping ought to be a recreation and a joy, the beginner is urged to make his start with representatives of a breed which he has found to approach his ideal of what a good hen ought to look like. He will be limited to some extent in his choice, as will be seen when the chapter on breeds is reached, but he may rest assured that almost any breed which he may decide upon will give him a plentiful supply of eggs if intelligently cared for. And in the long run, it is safe to say that the well-bred bird will prove more profitable than the scrub. Moreover, a flock of handsome, stylish hens is almost certain to receive better care than an assortment of mismatched fowls which make no appeal to the eye.

The time to begin keeping hens is just when the opportunity offers. There are several ways in which a beginning may be made. In the Fall, it is customary to buy a few pullets which were hatched fairly early in the Spring,—March if the breed chosen is about the size of Wyandottes or Plymouth

Rocks, a little later if smaller. It is well to have them in their new quarters by the first of October, in which month they may be expected to begin laying. If moved after the first eggs come, they are likely to cease laying for several weeks.

Pullets should be chosen in preference to older hens, because experience has shown that they lay better. No male bird will be needed; in fact, it is well not to have one in Winter. Eggs which have not been fertilized are preferred by discriminating buyers.

In February, a first-class, well-matured rooster of the same breed as the hens may be put with the flock. If the amateur lives in a closely settled neighborhood, he may find it advisable not to have a cock bird at any time, but to buy the eggs from which to hatch his chicks from a reliable breeder who has good stock.

After the first of the year, it may be well to buy a few two-year-old hens and a yearling rooster in order to secure eggs for hatching. Hens two years old are considered rather better to breed from than pullets, when mated with a younger male bird, and will cost no more — perhaps less. The beginner who starts at this time and in this way, though, should make it a point to buy his breeders from a man who has a flock of hens known to be good layers. It is the strain and not the breed that counts. Two men with adjoining plants may keep fowls of the same breed and yet one may get twice as many eggs as the other. He is the man who has selected his best laying hens year after year to breed from and so has perfected a prolific strain. The beginner who can get breeding stock from such a man will be fortunate. If he must, however, he can buy his birds at a poultry store in the city, but it will be the part of wisdom for him to admit his lack of expert knowledge and take an experienced poultry-loving friend along with him when he makes his choice. If he is able to buy his stock of a poultry keeper near home, he may not need to pay more than a dollar a head. Perhaps he will have to pay two dollars. If he seeks really fancy stock he will go to a professional breeder and pay according to his inclinations and the length of his purse.

When a start is made after the first of March and up to the middle of May, eggs for hatching may be purchased. More amateurs begin in this way than in any other. The eggs will cost from one dollar a setting up. In the country, it is often pos-

sible to get eggs from a farmer who has good utility stock for seventy-five cents for a setting of
thirteen eggs.   A motherly old hen to sit on the
eggs can be bought anywhere in the country for a
dollar.   It is wise, though, to make sure of the
hen before ordering the eggs.   Eggs may be sent
by express many hundred miles and hatch well,
although it is safer to buy them nearer home.

One should plan on setting one hundred eggs if
he wants a flock of twenty-five laying hens the
following Winter.   To hatch and raise fifty per cent.
will be a satisfactory record, and half of the chickens raised may be expected to be cockerels.   Of
course the latter may be served on the table or sold,
thus reducing the cost of rearing the little flock.
The twenty-five pullets left will be about the right
number for the average amateur and may be confined in a ten-by-twelve house.

There is another way of beginning in the Spring
and one which is growing in favor.   Day-old chicks
of most of the common breeds may now be purchased at from fifteen to twenty-five cents each, and
may safely be shipped for long distances, as newly
hatched chicks require no food for forty-eight hours,
due to the fact that just before they break their

shells, they absorb the yolks of the eggs, which provides them with an abundant supply of nourishment. Secured in this way, the chicks may be put under a broody hen at night or they may be raised in a brooder. This plan does away with all the work incidental to the care of sitting hens. Some amateurs find it desirable to hatch no chickens at all, but to renew their flock every year in this convenient manner. If it is too late to set eggs when the momentous decision to keep a few hens is finally made, this method of making a start is a very satisfactory short-cut.

The one draw-back to the purchase of day-old chicks is the fact that the buyer generally knows little about the stock from which the birds come. He may have bought into a good-laying strain and again he may not. Likewise, if he finds pleasure only in birds which are well-marked, he is participating in a lottery when he secures his chickens in this way. Custom-hatching has come as the solution of this problem. The amateur gets his eggs from whatever source he likes. They may come from his own flock or from that of a breeder known to have superior birds. The eggs are entrusted to the hatchery and the chicks turned over to the owner when they have pecked their way to freedom at the end of twenty-one days, a nominal fee compensating the man who owns the incubator for his work.

The perfection of mammoth incubators which have a capacity of several thousand eggs has given a tremendous impetus to this business of hatching chicks on a large scale and has made it possible for the man with a small flock to dispense entirely with sitting hens.   However he acquires his chickens, the beginner will do well to bear in mind the point already brought out, that he will inevitably lose some in the course of the brooding period, and half at least of the number raised may reasonably be set down as cockerels.

I have said that twenty-five hens is a good number for the amateur, but there is no reason why the flock should not be smaller.   It is a reassuring fact that the small flocks lay the largest percentage of eggs.   There are little portable houses costing about ten dollars in which eight hens can be carried through a Winter and that number of hens will keep a small family in eggs under proper conditions.

It is not necessary to have an outside yard for the poultry to run in, although such a yard is preferable because less work is required than when the hens are confined all the time. It is perfectly feasible, however, to keep a small flock housed all the year round if the houses are of the open-front or fresh-air type and not placed in too hot a situation. As a rule, though, it is not advisable to breed from hens kept in such close confinement, so amateurs who use this plan, sell all their laying hens in the Summer and buy well-grown pullets in the Fall, unless they care to purchase eggs or day-old chicks in the Spring. The custom of stocking up in the Fall with pullets just ready to lay and selling off the hens in the course of the Summer as they cease to lay makes poultry keeping a very simple matter and solves the problem for the would-be amateur who hesitates to begin because he has no time to give sitting hens or an incubator and no place in which to rear young chickens.

It must be said, though, that growing the chicks is one of the most fascinating and interesting phases of poultry keeping to the amateur who is a genuine, seasoned enthusiast. There are professional men who find their chief relaxation in the Spring of the year, when business cares weigh heavily and vacation days are yet far away, in the care and oversight of a little flock of chickens — sometimes Bantams, which have many admirers among busy lawyers, doctors and ministers.

The amateur who is wise will not waste a lot of money on elaborate houses for his birds. It is a curious and amusing fact that almost every poultry keeper has a pet theory about the best kind of hen houses. The theory held to-day, however, may be quite different from the one held yesterday, so that it is well to allow for alterations and remodeling. Furthermore, simple houses are by all means the most practicable. It is well enough to spend as much money as one can afford in giving the poultry house an architectural finish to make it harmonize with the other buildings on the estate, but there should be no frills within doors and every fixture should be detachable, so that it may be removed and cleaned.

If the beginner must count the cost, he ought not to spend on his poultry house more than a dollar for every hen which is to be confined there. This is a good basis for the prospective poultry keeper to lay his plans on when he is figuring the expense of the venture.

Of course, poultry houses in suburban sections which must occupy somewhat conspicuous situations will need more attention to external appearances than those in less prominent locations. There are always one's neighbors to consider and it is altogether selfish to put up a building which will prove an eye-sore to anyone. Then, too, there is a false economy. The poultry house should be well and strongly built for the sake of the birds that are to occupy it. The owner will find more satisfaction in caring for his fowls if they are kept in neat, attractive quarters.

It is a mistake for anybody who has not had experience in keeping a large flock of poultry to begin on a large scale. Failure is almost sure to follow. Some amateurs expect to make a considerable profit from a small flock, and that, too, the first year. Of course, they don't do anything of the kind. Keeping hens is not a royal road to wealth. One dollar per year per hen is considered a fair profit. Some men and women make more, but even though the profit be twice that amount and the flock number a thousand, which means hard and constant work for one person, it will be seen that the amount of money to be made is not great.

Keeping a few hens for recreation and to supply the family table with fresh eggs and chicken is, however, to be highly commended to every man and woman who has the time and opportunity to give the birds the very little care they demand. The scraps from the family table will go far toward supplying the rations of a small flock and the satisfaction of eating eggs laid the day previous may well be imagined by those unfortunate people who are never certain whether their breakfast egg was laid a month or a year before it appeared on the table.

CHAPTER II

SELECTING A BREED TO KEEP

SPEAKING broadly, the amateur will be wise to select the breed and variety which makes the strongest appeal to him. He will have a wide choice, both as to size and markings. Yet there are several points aside from appearance which demand consideration. For example, some breeds lay pure white eggs, which in certain quarters is felt to be a distinct disadvantage. Leghorns are bred in comparatively small numbers in New England because the people living there have a strong prediliction for brown eggs, such as are produced by Plymouth Rocks and Rhode Island Reds. New York epicures, on the other hand, are willing to pay a premium for pure white eggs, and so Leghorns, the most famous breed laying eggs of that color, are bred by most of the poultrymen catering to that market. Following the fashion, the majority of New England amateurs keep hens of breeds which lay brown eggs, while in New York, New Jersey and Pennsylvania the white egg hens are more popular.

Of course, this matter of color preference is merely a notion, fostered by custom. There is absolutely no difference in the quality of a brown-shell egg and that of one with a white shell. On the Pacific coast an effort is being made to lead people away from their unreasoning prejudice for white eggs, so that the amateur will be encouraged to keep the general purpose breeds, which lay brown eggs. In most sections, though, the professional poultrymen who have a white-egg market are well satisfied, for the White Leghorn is the nearest approach to an egg-producing machine which has yet been developed.

There are other points to be considered, too. Hens like the Leghorns and Anconas, which are marvelous layers, are very small and so of little value as table fowl. Moreover, they are non-sitters, which means that it is impossible to hatch eggs under them or to raise chickens with them, making it necessary for the amateur to keep a few hens of another breed or to rely upon incubators, if he raises his own birds. Also, these light hens are high flyers, so that more fencing is needed than for the heavier breeds, and they are not so contented when closely confined, while they always remain rather wild and cannot be petted like the representatives of the larger breeds.

Some of these qualities, however, commend them to certain amateurs. The fact that they never sit is much in their favor if an incubator is to be used or if no chickens are to be hatched or raised, or if day-old chicks are to be purchased and reared in a brooder. Broody hens are a nuisance under such circumstances. The fact that these light breeds eat much less than the larger ones is distinctly in their favor and the difference in the amount of food consumed by a Leghorn and a Rhode Island Red, for instance, is surprisingly large, especially when it is considered that the smaller hen will lay the most eggs, as a rule, and that the eggs are often as large. The author has been keeping a pen of Anconas side by side with a pen of Reds, and the eggs of the former have averaged notably larger.

It should be said, though, that the matter of strain enters into this proposition as well as into the number of eggs produced. Some strains lay much larger eggs than those of other strains of the same breed, for some breeders make a point of breeding

for large eggs. One other point in favor of the
light breeds is found in the fact that more of them
may be kept in a house, four square feet to each
bird being sufficient, while at least five square feet
are required if the larger breeds are to be comfortable.

The matter of color is also to be considered. If
kept in a town where the air is filled with smoke
or where the soil is highly colored and heavy, white
fowls are not easily kept in a presentable condition
and black ones or those with dark-colored feathers
are to be preferred. White fowls in a clean city
look especially handsome when allowed to run on
the lawn and may be preferred for their ornamental
value. But if the chickens are given a wide range,
those which are white will become shining marks
for hawks.

The poultry of to-day is divided into several distinct classes, known as Asiatic, American, Mediterranean, English, French, in addition to which there
are Games, Bantams, and a few miscellaneous
breeds. Fowls of the American and Mediterranean
breeds are those most commonly raised in this country. The American class comprises the great utility
breeds like the Plymouth Rocks, Wyandottes, and

Rhode Island Reds, which are prolific layers of large brown eggs and which are heavy enough to dress well for the table. These are the breeds usually found on the farms where pure-bred fowls of any kind are found. They are hardy, easy to care for and tame; they are good sitters and make good mothers.

Most breeds are divided into several varieties and some into many like the Wyandottes, of which there are white, buff, silver, golden, silver-penciled, partridge and Columbian. There are barred, white, buff and Columbian Plymouth Rocks, but the Rhode Island Reds are confined to one variety, except that some have single and some rose combs. They originated in the state from which the breed takes its name and among professional poultrymen who were seeking a superior all-round fowl. The Red has grown amazingly in popularity and is now closely crowding the Barred Plymouth Rock, which for years was the one breed seen everywhere. A Rhode Island White has now appeared.

The one objection to the Reds from the point of view of the amateur who has a liking for handsome hens is the difficulty which is experienced in getting uniformity in coloring. There are many variations from the reddish buff desired, but by close culling, as well as careful breeding, it is possible to get a flock in which the color of all the birds is practically the same, and such a flock must delight the eye of any amateur.

All the breeds in the American class have yellow flesh and yellow legs, which is to be expected of birds bred in this country, where much weight is given to these points by the buyers and sellers of dressed poultry.

The fact that the flesh is white is a serious drawback to the popularity of the Black Langshan, which is the only representative of the Asiatic class commended to the consideration of the amateur. The Langshan is smaller and more active than the other Asiatics, but has feathered legs like the others. The cock is a regal-looking bird and there are few handsomer or more stylish fowls than a Black Langshan hen. A very low fence will confine the Asiatics, which are very quiet, slow-moving birds. Years ago the Light Brahmas were immensely popular and many old-time admirers of this breed have now taken up the Columbian Wyandotte, an American breed which has practically the same markings, but which is smaller and with clean legs.

Of the Mediterranean class, the Leghorns easily take the lead in popularity. Probably there are more heavy egg-laying strains of White Leghorns than of any other breed. There are rose-comb as well as single-comb white and brown Leghorns, the rose-comb varieties often being recommended for very cold climates, as the rose combs are not frosted as easily as the longer single combs. The single-comb white Leghorn is generally admitted to be without a peer as an egg producer and the eggs are valued so highly that the best trade in New York often specifies them.

Like all members of this class, the Leghorns are non-sitters. They have yellow flesh and legs but are too small to be considered as table fowls, and so are not so popular in the country and in the yards of many amateurs as all-round fowls like the Rocks and Reds. The meat, however, what there is of it, is especially fine-grained and sweet and the chickens make fair broilers. Mature Leghorns are hard to fatten, though, and must be set down as belonging to a strictly egg-laying variety.

The two varieties of Minorcas, white and black, are somewhat larger than Leghorns. They, too, are prolific layers of large, white eggs, but they have white skins and dark shanks, which puts them at a disadvantage on this account. The Black Minorcas are the more common.

Anconas have achieved unexpected popularity within the past few years, especially in the West. They have most of the characteristics of the Leghorn, being small, active and exceedingly prolific, while their eggs are large, white and well shaped. Anconas are black, except that every fifth feather has a white tip, giving the birds a very pleasing mottled appearance, which doubtless has helped to win the favor of breeders with an eye for striking and attractive markings.

The Black Spanish and the Andalusians are satisfactory breeds for egg production but are bred mostly by fanciers, being, for one reason or another, not in great favor with the majority of practical poultry keepers. Their markings are odd and interesting. The Andalusians are especially curious, as their feathers have a bluish tinge. Interest in blue fowls has increased and there are blues in other classes.

Of the English class, the Orpingtons in white, buff and black varieties are enjoying considerable prestige in this country, surprisingly large entries being made at some of the Eastern shows. They are splendid all-purpose fowls and so aristocratic in manner that they make a strong appeal to many beginners. They are a trifle larger than the American breeds and doubtless would speedily come into great favor if it were not for the fact that their flesh is white. They lay brown-shelled eggs and produce them in generous numbers. The meat is excellent and a bird which is well cared for carries a lot of it, especially on the breast, so that for home use, the black Orpington is highly satisfactory, particularly when it is desirable to have a breed with plumage which will not show stain or dirt.

The Houdan is the French fowl best known in this country and is an excellent breed for the amateur, as the birds are tame, unusually attractive, contented in confinement and good layers. The eggs are white and the meat is too dark for market poultry, but exceptionally fine in texture and flavor. There is not so much meat on the carcass as is furnished by birds of the American and other larger breeds, but more than the Leghorns offer. The flesh of Houdans is highly prized in France, where dark-meated fowls have the preference. The Houdans have crests or top-knots and are mottled black and white in color. They are hardy and non-sitters. It is a characteristic of Houdan eggs that they are unusually fertile, while the chicks feather out rapidly and come to maturity very early. Houdans have one peculiarity in the form of a fifth toe, like a very old English breed, the Dorking. The Hamburgs and the Polish in several varieties comprise two classes very popular with fanciers.

The Games and Bantams in various breeds are freely bred by fanciers and afford much pleasure to their owners. Many of the Bantams are pocket editions of larger breeds, as, for example, the white, black and buff Cochins and the Light and Dark Brahmas. Probably Buff Cochin Bantams are bred more widely than any of the others, but all of those named, as well as several other kinds, will afford their owners no little pleasure. Most of these Bantams lay a considerable number of eggs, especially in the Spring, and the eggs, although small, are rich. Three are equivalent, as a rule, to two eggs from ordinary hens. The breeds mentioned are large enough to dress for the table. Bantams can hardly be surpassed as pets for children, when the latter are old enough to care for them. A packing box is large enough to house a small flock and the amount of food consumed is very small.

Below is a list of the more common breeds of poultry, with notes on some of their marked characteristics:

ASIATIC CLASS

Name	Weight lbs.		Sitters	Color of eggs	Color of flesh	Varieties
	Cocks	Hens				
Light Brahmas...12		9½	Yes	Brown	Yellow	
Dark " ...11		8½	Yes	Brown	Yellow	
Cochin 11		9½	Yes	Brown	Yellow	Buff, partridge, white, black.
Langshan 9½		7½	Yes	Brown	White	Black, white.

AMERICAN CLASS

Plymouth Rocks.. 9½		7½	Yes	Brown	Yellow	Barred, white, buff, silver, partridge, Columbian.
Wyandottes 8½		6½	Yes	Brown	Yellow	White, buff, silver, Columbian, golden, partridge, black, silver penciled.
Rhode Island Reds 8½		6½	Yes	Brown	Yellow	Single and rose comb.

MEDITERRANEAN CLASS

Leghorns .... *		*	No	White	Yellow	Single and rose-comb white, single and rose-comb brown, single and rose comb buff, black, silver.
Minorcas .... 9‡		7½‡	No	White	White	Single and rose comb black, single comb white.
Anconas .... *		*	No	White	Yellow	

ENGLISH CLASS

| Orpingtons ......10 | | 8 | Yes | Brown | White | White, black, buff. |

FRENCH CLASS

| Houdans .... 7½ | | 6½ | No | White | White | |

* Have no standard weights.

‡ Weight of single comb black.   The other varieties are a pound lighter

It seems almost impossible that the almost innumerable breeds and varieties of poultry the world over should all have sprung from one common parentage, and yet it was Darwin's theory that the jungle fowl of India was the progenitor of all known kinds. As the fowls were scattered over the earth, they were developed in many different ways. The poultry of China and Japan is quite different in appearance and characteristics from that of this country. In France breeds with white flesh and dark legs have been perpetuated, because the Frenchman likes that kind. In this country breeds with yellow flesh and yellow shanks have been made by crossing older breeds and carefully fixing the type in order to satisfy the Yankee prejudice.

The whole subject is full of interest, and it becomes plain that the amateur has wide latitude in making his selection. Almost every breed has something to recommend it, and no breed is entirely free of faults, so that it is not wise to skip from one to another if the first breed chosen does not prove as satisfactory as expected. The better plan is to give that breed a little more study in order to learn if the fault is not with the keeper instead of with the hens.   And finally, it is not the part of wisdom to keep more than one breed at a time, when one first begins his poultry work.

CHAPTER III

THE KIND OF HOUSE TO BUILD

WHAT a satisfactory thing it would be if one could draw a plan and say, " That is the best kind of poultry house for the amateur to build." But what a riotous time the man who should attempt that sort of thing would have! Poultry experts differ no less radically than doctors, and probably more time has been devoted to planning poultry houses than to designing churches.

Of course the writer has his personal pet theory about poultry house construction, but he is not parading it, for it may change in the future as it has in the past. Few poultry keepers indeed would construct to-day the kind of house they would have built ten years ago. A distinct advance was made when the discovery was announced and proved to be true that poultry would thrive better in cold but dry houses in which there was an abundance of fresh air at all times than in very warm houses in which the ventilation was poor with the result that moisture

collected on the walls and made the houses damp.
No longer are houses built with double walls packed
with sawdust and with tightly closed windows filling
the front.   Many poultrymen have gone so far as to
omit all glass, either substituting muslin curtains or
making their houses entirely open in front.

The conservatively radical house, if the expression may be permitted, has much muslin and a little
glass, the latter admitting light on very stormy days
when it is advisable to have the muslin covered
frames closed, and also, if properly arranged, allowing sunlight to strike directly on the floor at the
front of the pen early in the morning, something
quite desirable in cold weather, when the sun is slow
in rising.

There is no better type of house, in the opinion
of many experts, than one which has a long hori-
zontal opening, the bottom of which is two feet
above the floor, with a window of glass under it or
a taller window at the end, placed upright and with
the bottom close to the floor.   The long opening
should be fitted with a muslin-covered frame, which
may be hinged at the top and hooked up out of the
way when not in use.   Of course the opening should
also be covered with poultry netting, and it is well

to use one-inch mesh netting, so that sparrows will be kept out, as these birds often steal much more grain than they are entitled to, for any good they do. A house of this type, in use at the Perdue University Agricultural Experiment Station, Lafayette, Ind., is illustrated. The bottom of the window is low enough so that there is direct sunlight on the floor of the house early in the morning during the winter months and the opening is made high enough from the floor so that the wind will not blow directly on the birds. In a house of this character almost every interior part is reached by sunshine at some time in the day.

In these fresh-air houses the muslin curtains are not to be used except when the mercury is unusually low or when the rain is being driven in. Their only disadvantage is the fact that they collect a great deal of dust and need to be cleaned frequently, so that light may penetrate them. It is a good plan to have them on pin hinges, so that in Summer they may be taken down by drawing the pins, and put away until cold weather comes again.

It is a common practice in very cold climates to have a second set of curtains in front of the perches, to be dropped when necessary. Such curtains should

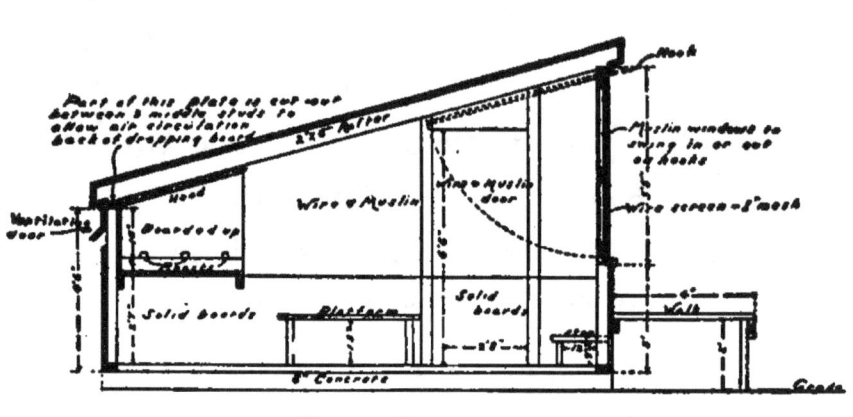

Cross Section

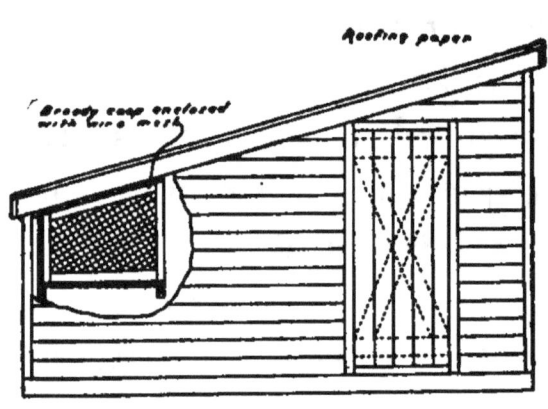

Side Elevation

Side elevation and cross section of the Perdue University house. The ventilating device at the rear is well planned and the broody coop of poultry netting is a valuable feature

not be used unless the temperature is much below freezing and burlap or old bagging run on a wire will answer as well as muslin tacked to a frame. It should strike the front of the dropping boards or hang to the floor.   The great advantage in cloth is that it admits air freely but without drafts.   Experiments have shown that buildings where muslin has been used at the windows have been only a degree or two colder than when glass was used, for glass radiates cold.

Poultry houses in the South need no protection at the windows and the type which has the entire front open gives full satisfaction.   Indeed, open-front houses are being largely used in the most northerly states and many poultry keepers are enthusiastic in praise of them.   They certainly simplify the keeping of poultry, for there are neither windows nor curtains to look after, the front of the house being entirely open except that it is covered with poultry netting to keep the birds in and intruders out.   In some cases a canvas curtain is dropped over the front when necessary to keep out snow or a beating rain, and occasionally curtains are used in front of the perches on extremely cold nights, but the average owner of an open-front house

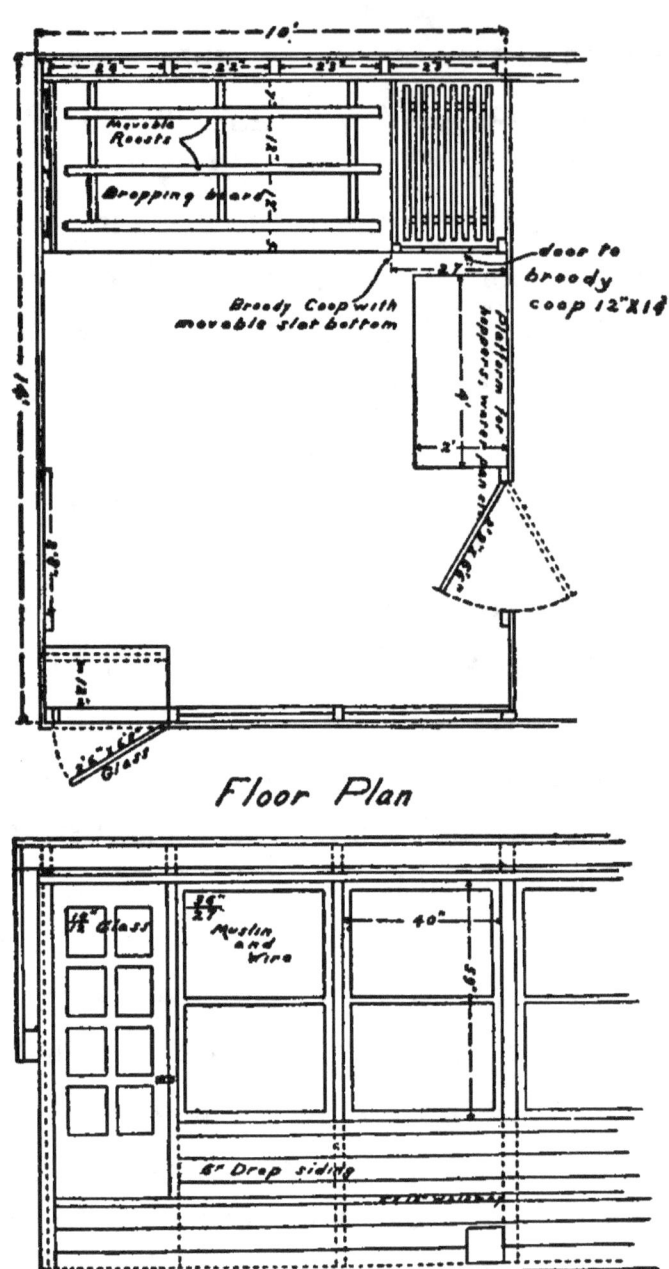

Floor Plan

Front Elevation

Details of the Perdue University house. It can be made with
one pen as well as with two

goes to the extreme of giving the hens no additional protection at all. Strange as it may seem, too, this method gives excellent success in scores of instances. The hens show no evidences of suffering from the cold, seldom have frosted combs, lay steadily through the coldest weather and are obviously in the pink of condition.

Several things need to be considered, however, in building a house of this kind, or the results will not be so satisfactory. In the first place, a deep house is required. If the house is only eight or ten feet deep, the birds will be sure to suffer. Then, there must be an opening on one side only or the house will be too cold. When the pen is a deep one and open only at the front, the wind meets an air cushion when it strikes this opening, for the air within cannot be forced through. Such a house should be perfectly tight as to walls, and may have a glass window on the west side, if deemed desirable in order to get the afternoon sun. This is an advantage in a house of the type which is rather low in front. The well-known Tolman house is an example of this type. It is a long house, with a double-pitched roof, the ridge being about two-thirds the distance from the front, so that there is a long, gradual slope in front and a sharper pitch at the rear. The hens roost at the back of the house and the air cushion keeps the room so warm that Mr. Tolman finds curtains unnecessary, although he lives near the coast in Massachusetts.

On the famous Hayward farm at Hancock, N. H., " A " shaped houses are used, each accommodating a dozen birds. The front is covered only with poultry wire and no protection in the form of curtains is given.

Houses of a type which is the acme of simplicity are in common use in Southern New England, many of them on the farms of men who make poultry keeping a business. They have a double pitch roof, seven feet at the ridge and about four at the walls. The length of the house is about sixteen feet and the width about eight feet; one end, facing the south, is open. The fowls roost at the opposite end, of course, and thrive in a house of this character, which fact satisfies the men who build them that the plan is a good one. It is needless to say that such a house is inexpensive; and it will provide quarters for thirty or more fowls.

Unless one is prepared to brush off his muslin curtains every other day and can resist the impulse to close them when the weather gets a little chilly, he had better consider the open-front house carefully, always bearing in mind that he can drop burlap in front of the perches if he finds it necessary. Curtains that are not frequently cleaned, soon become clogged with dust, so that they admit little more air than a board. Then, of course, their chief merit, that of providing ventilation, has departed.

The tendency in the direction of open-front houses has brought about another change, in the form of deeper houses than have been common, the reason of which has already been explained. Houses twenty feet deep or more are now being planned, but in such a house the sunlight will not reach the roosting quarters unless a shed roof is used and raised to a quite impracticable height in front. As direct sunlight is most desirable for sanitary reasons, windows in the roof have been devised. Sometimes the semi-monitor type is adopted.

Such a house is in use at the New Jersey Experiment Station, and the type is warmly commended to farmers in particular by practical men. More than ordinary interest is being shown in this house because it is a radical departure from the type which has been generally recommended of late years. It is 40x20 feet, there being two pens each twenty feet square, which might be made the size of a smaller house. If this house were but ten feet deep, it would have to be eighty feet long to give the same number of square feet. A square house is more economical to build than a long and narrow one, for it requires less material. The arrangement of the windows at the top of this house makes it light in Winter and helps to keep it cool in Summer. A house that is only four feet high at the rear, where the fowls roost, is warmer at night than one which is higher, but this is a distinct disadvantage in Summer, and poultry often suffer as much from a high as from a low temperature.

Another way of lighting and ventilating deep houses and a somewhat cheaper one, is to have windows placed sky-light fashion in the front slope of a double-pitched house. The plan has sometimes been tried but reported a poor one because water came in around the windows. This trouble may easily be overcome by attaching zinc strips to all sides of the windows, so that they will come outside the sashes when the windows are closed. Of course, it may be necessary occasionally in Winter to clean the snow from the windows, but as a rule the

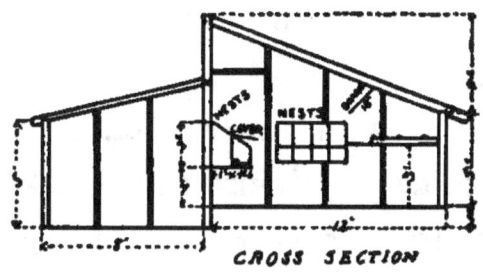

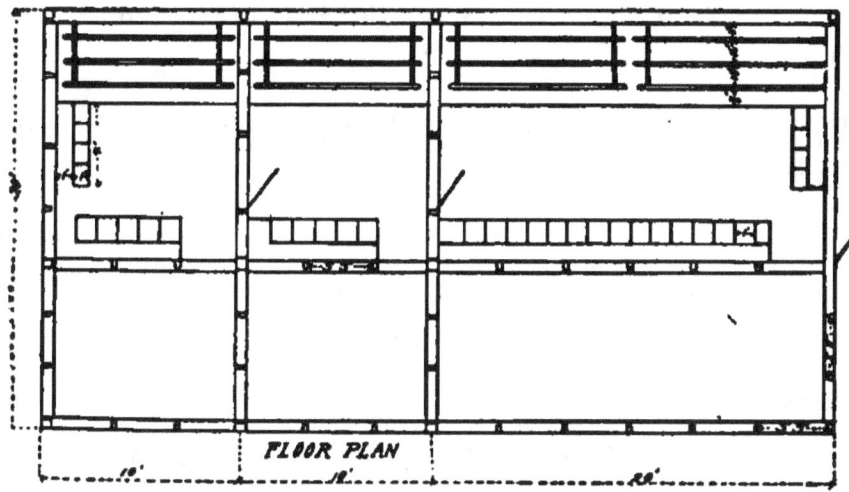

Details of the Clark semi-monitor, open-front house at the
New Jersey State Experiment Station. A photograph of
this house is shown on page facing 36

slightly higher temperature resulting from the warmer air in the house will cause the snow on the glass to melt and run off.

Speaking broadly, the most satisfactory house for the average beginner is one the dimensions of which are about 10x12, with a shed roof seven feet high in front and four feet at the rear. Such a house will accommodate from twenty-five to thirty fowls comfortably.

Renters look with favor upon portable houses, for if they move, the house can be taken apart and carried along to the new place of residence. Houses large enough to care for a dozen or fifteen hens cost from twenty to thirty dollars. They have dirt floors but are tight and well made, being high enough so that the attendant may work in them with ease.

Another form of portable house is much smaller and not high enough for a man to enter, but made to accommodate eight or ten hens. The roof is so arranged that it may be raised from the rear in order to facilitate cleaning the dropping board and any other work inside the house which may be required. A house of this kind with a scratching shed attached costs about ten dollars and a dollar or two more will be needed to pay for the roofing paper needed to cover it so that it will be suitable for use in cold weather. Houses of this type have been tested and proved wholly practical. Muslin is used in place of glass, and being low, the house is quite warm enough.

Unless the question of ornamentation or architectural finish figures, a thoroughly satisfactory house, as has been stated, need not cost more than one dollar for each hen which is to occupy it. Considered purely from the financial standpoint, a poultry house should cost no more. Many practical houses in which paying flocks are kept cost less. Many houses, on the other hand, which have cost much more have proved a failure. The writer once visited an amateur's plant on which there was an expensive house plastered on the inside and with a stove to keep it warm. It was large enough for many score of fowls, but only a few lonesome-looking birds were to be seen wandering about.

Single boarded walls covered with roofing paper are sufficient. Paper is also to be preferred for the roof, as a rule, because it wears well when the roof slopes toward the north and because the pitch of the roof need not be as great as when shingles are employed. The principal point in building the poultry house is to have it dry and perfectly tight. With matched boards and a good quality of patent roofing, this purpose is accomplished.

The floor may be of earth, boards or concrete, depending largely upon circumstances. In locations where water seeps away quickly, earth floors answer as well as any, but they should be built up at least a foot higher than the surface of the ground outside. The earth should be packed solidly and several inches of gravel thrown on top, to be removed and renewed once or twice a year. If rats bother, inch-mesh poultry wire may be laid several inches under the surface and made fast to the foundation. Another plan is to shovel out the earth around the outside of the house to the depth of a foot and the same distance from the foundation, and to place poultry wire so that it will extend down a foot into the ground and then a foot away from the house. This will prevent the rats digging in, as they always work close to the foundation.

If the house is on a side hill or elevated from the ground, board floors may be used; but they should not be laid over damp ground with the expectation of keeping the house dry. An earth floor well elevated is better under such conditions. If, however, several inches of stones or cinders are placed under the boards, the result will be satisfactory. When there is an opening under the house, the floor will be cold unless double boarded. Of course, a layer of building paper between the boards will help. Rats may be kept from coming through a double-boarded floor by putting wire with a close mesh between the boards.

Cement floors may be good or bad. If laid on the ground, moisture is very likely to rise through the cement. Some experiments of this kind have been most unsatisfactory. If the earth is dug out and replaced with a layer of stones or cinders a foot deep, this trouble will be avoided. Cement floors are cold to the feet and should be covered with several inches of sand, with a deep litter above. If well made, a cement floor is durable and rat-proof and the cost is not great. Whatever kind of floor is used, it should be several inches higher than the ground outside, that there may be no possibility of water collecting on it.

The foundation walls for the permanent poultry house may be made of cement to advantage, although stones are often used. A cement wall is easily made by digging a trench to a point below the frost line, lining it with boards and filling in with layers of small stones and cement. Boards fastened to stakes will hold the cement above ground. Such a wall keeps out the rats.

Cedar or hemlock posts set into the ground deeply make good supports for the small poultry house. In England, a favorite plan is to have the house on runners, so that it may be moved from place to place about the grounds, and sometimes wheels are used. The use of small houses which can be moved helps to make the work light, if there is plenty of room, for they can be moved instead of being cleaned out.

It will be seen from all this, that the question of poultry house construction is not as simple as the novice might be led to believe. However, the fact remains that for the beginner there is nothing better than a simply made, shed-roof house with a combination of muslin and glass. Even this assertion may be disputed, too, so that it is not made dogmatically.

The furnishings of the poultry house should be as simple as the house itself. Lice the poultryman always has with him. By that is meant that there will always be a considerable number in the house, but if care is not exercised, he will have them literally with him, for they will swarm over his person every time he enters. Nothing will make the beginner more thoroughly disgusted with the poultry business than this experience. In order, therefore, that every part of the house may be cleaned, the furnishings should be detachable as well as simple, so that they may frequently be taken out and given an application of kerosene or some other liquid for which lice and mites have no liking.

The best perches are made of 2x3 scantling set on edge and with the top corners rounded. If there is more than one perch they should be the same height and not arranged stepladder fashion, for then every hen will seek the highest round and there will be nightly confusion and quarreling. The perches should be not more than three feet above the floor and rest in a slotted board at each end, so that they will not need nailing. If perches are made very high, heavy fowls often suffer injury when jumping down.

The dropping boards may be some twelve inches lower and should fit tightly against the wall at the back. It is well to have the boards cleated together so that they may be taken out at any time. A piece of scantling nailed to the wall at each end will support them. Some amateurs have adopted the plan of supporting the dropping boards on wooden horses, so that the wall may be kept free even of braces, thus lessening the number of cracks in which insects can hide.

Other poultry keepers have abandoned dropping boards entirely, but set a board upright in the floor to confine the droppings to the rear of the house. Practical poultry keepers find that this plan greatly lightens the labor as there are no dropping boards to clean, and that by keeping a little earth, peat or old litter under the perches, there is no trouble from odors if the accumulation is removed every month or two. Probably the average amateur will consider it neater to use dropping boards and he can perhaps handle the manure to better advantage. Still, they have no place in some of the portable houses now being exploited as especially adapted to the needs of the amateur. If boards are used, coal ashes, peat or dry earth should be dusted over them to act as an absorbent. Wood ashes should not be used if it is desired to retain the fertilizing qualities of the manure.

Nests twelve inches square and as high are serviceable and may be hung to the walls with screw eyes. Often the nests are put under the dropping boards and entered from the rear, but such an arrangement makes it difficult to reach any eggs which may be dropped in the alley back of the nests as well as decreasing the floor space.

Dark nests are not required, but there is less danger of the hens acquiring the egg-eating habit if they are used. One amateur makes boxes three feet long with an opening at one end and a hinged cover. He finds that there are fewer broken eggs and less quarreling when several hens crowd into such a nest at the same time than when they are given individual nests.

The covers should always have a sharp pitch so that the hens cannot roost on them. In many cases, egg or orange crates turned on their sides are proving very satisfactory nests. Nest eggs are of value in teaching or rather coaxing the hens to lay in the nests instead of in the corner.

It is a good plan to have a platform for the feed and water dishes, so that the hens cannot scratch litter into them. A galvanized water pail is as satisfactory as a patent fountain when this plan is followed.

A coop for breaking up broody hens is desirable, if hens of the breeds which have the sitting habit are kept. It may be made easily enough by using laths for the sides and the bottom and may be hung on the wall. It should be large enough for the hen to move about in comfortably, but she will soon lose her broodiness in such quarters, finding it very unpleasant to sit on slats with wide spaces between. Some people use poultry wire instead of laths. There is no need of starving a sitting hen or abusing her in any other way. Indeed, bad treatment will make her slower in beginning to lay again. Our grandfathers who used to duck sitting hens in the watering trough and slap them on the barn floor simply did not know any better. There has been progress in the poultry world.

Chapter IV

FEEDING A LITTLE FLOCK

FEEDING a small flock of hens may be a complicated or a very simple matter. Observation leads to the belief that the man who adopts a simple but intelligent plan will be just as successful as the one whose methods are more complex, and with much less effort. Many amateurs coddle their birds too much and overfeed them. Even people who keep but a few hens like to have them show a profit and it must be remembered that profit represents the difference between receipts and expenses. Some flocks which lay heavily fail to pay as well as others which produce fewer eggs but at less cost. When hired help is employed, the labor required becomes an important item. The fact is that hens do not need a great amount of fussing over. This statement is true, also, of chickens. The man who goes out with a lantern at 10 o'clock at night to give his brooder chickens a final feeding is not a wise poultry keeper.

So much has been written about egg-producing feeds and growing feeds and balanced rations that the bewilderment of the beginner who tries to comprehend it all can hardly be wondered at. It is needless to puzzle long over balanced rations. Place a trough of mixed grains before a flock of hens, and no matter how carefully the proportions have been balanced, the birds will unbalance them in about three minutes. That is to say, each fowl picks out the particular kind of grain which she likes best — and some poultry have very decided preferences.

Good sound grain in variety, with a mixture of ground grains served as a mash, a certain amount of meat in some form and green food in abundance will fill all the requirements. It must be remembered that hens forced for eggs do not make good breeding stock, besides demanding more care. The amateur's flock need not be fed oftener than twice a day, although it often is convenient to give the hens the scraps from the family table at noon. On one famous egg-producing plant the hens are fed at least five times a day. But those hens are forced. They might be termed specialized hens.

The grains to use are corn, oats, wheat, barley

and Kaffir corn. Corn, oats and wheat are the grains to be depended upon month in and month out. The others are fed to give variety, but really are not necessary. Many warnings against the excessive use of corn have been sounded, so many, indeed, that some breeders have come to be almost afraid of this grain. Yet corn is the best poultry food there is and the danger that it will make the fowls too fat to lay is a bugaboo to which little attention need be paid, so long as pullets and one-year-old hens are kept — and nothing older ought to be found in the poultry yard. A fat hen will lay better than a thin one, anyway, and pullets with ordinary freedom and an average amount of mixed feed will seldom become unduly afflicted with avoirdupois.

Corn, then, may well constitute one-third of the ration the year around, and in Winter no harm will be done if half the scratch ration is corn. Of course this grain is heating, and in extremely warm weather the quantity may be reduced to a very small amount for the time being.

Many amateurs get satisfactory results feeding equal parts of cracked corn, oats and wheat. Cracked corn is better than whole corn simply because it makes the hen work harder to fill her crop, and exercise is important. If it is necessary to defer the afternoon feeding to only a short time before darkness falls, whole corn should be given, so that it will be cleaned up quickly. The different grains may be mixed and fed together or divided in any way one prefers. The author's plan is to feed oats and wheat, or barley and wheat in the morning and cracked corn at night, giving the hens at least an hour in which to clean up the last feeding before they feel inclined to seek their perches. Clipped oats are especially desirable.

Most families have many table scraps which may be fed the hens to advantage. They may be run through a meat grinder or made into a mash by soaking them and mixing in a little bran or meal. Some amateurs practice the plan of keeping a kettle on the back of the kitchen range into which go all the scraps as fast as they accumulate. In this way they get softened and cooked and may be made into a mash as needed. Feeding this mash at noon breaks the monotony of the day for the hens, but it may be given in the morning or at night just as well. Experience has shown that if a mash be fed about 4 o'clock in the afternoon, the hens will still eat a large amount of corn if this grain be given them at 5 o'clock.

It is always well to supplement a mash ration fed in the afternoon with grain, so that the birds' crops will not be emptied too quickly. Whole or cracked corn is not digested nearly as quickly as mash. It is a mistake to feed so much mash in the morning that the hens will sit around for an hour or two before beginning to scratch in the litter. This is especially true in Winter, for then the fowls get thoroughly chilled.

There has been much discussion over the relative merits of wet and dry mash. By a wet mash is not meant a mash which is really wet, but one which has been moistened with water, milk or some other liquid, yet which is dry enough to crumble in the hand when squeezed. The sloppy mash of our grandmother's day is taboo among all up-to-date poultry keepers, amateur or otherwise.

Dry mash means simply a mixture of ground grains which has not been moistened at all. Dry beef scraps, meat meal or shredded fish and dry alfalfa or clover may be added. The dictionary says that a mash is a soft or pulpy mass, which simply goes to show that the dictionary maker was not a poultry keeper. Dry mash is a term invented by poultrymen to contrast meal and various grains ground finely but fed dry, with the same combinations moistened with milk or water. Amateurs are sometimes puzzled by the expression, just as they are deceived into expecting a wet mash to be really wet, when it is, when properly made, only a crumbly mass.

It is probably true that a few more eggs are secured when a wet mash is given once a day, but against this advantage must be set the extra work required plus the fact that the eggs are likely to be less fertile. This does not refer to the mash which may be made with table scraps, but one mixed regularly of various ground feeds, meat and alfalfa. If such a mash is given, of course the table scraps should be mixed with it, but otherwise the latter are to be considered a supplementary feeding, although assisting in cutting down the grain bill. Of course, if the family is a large one, there may be enough scraps to make one full feeding, with the addition of some bran and meal, and it would be wasteful in such a case not to make the most of it.

Probably the average amateur will find using dry mash in hoppers the most satisfactory plan. He will save labor and be reasonably certain of good results if he uses one of the commercial laying mashes, which can be bought at seed and poultry supply stores anywhere and the best of which contain a well-balanced mixture of many grains with cut alfalfa and beef scraps added. Placed in a hopper where the fowls will have access to it at all times, the amateur never need fear that his hens are being underfed or be conscience-smitten if he is obliged to skip a feeding of whole grain. In fact, if it is found that only a little of this mash is being eaten, the amount of grain fed may be reduced in order to drive the hens to the hoppers, for the birds which eat liberally of the mash will usually be the ones to lay best.

The tendency is to feed too much grain, and the amateur is especially likely to err in this direction. It goes without saying that the grain should be fed in a litter several inches deep, and it is wise to poke about in this litter with the foot occasionally to see if there is any grain on the floor. If grain is found there, the fact may be taken as indicating that the birds are receiving at least all they need, and it is a good plan to purposely miss a feeding once a week so that the hens will be forced to scratch industriously in order to get the grain which has been accumulating from day to day. On one commercial plant, the evening meal on Sunday is always omitted.

There are hoppers in great variety and at low prices. The kind which may be easily closed at night and which are made of metal are particularly desirable, because they prevent loss of grain from the depredations of rats and mice. A hopper may be easily made at home, however, by using a soap box, and a few moments' study of a ready-made hopper at the store will be enough to suggest to a man handy with tools how to go about the job. The amateur with but little time at his disposal will do well to use a hopper large enough to hold sufficient mash for a week or more.

If one prefers to mix his own mash, he may adopt one of several formulas. A simple one is: Three parts bran, two parts ground oats, two parts middling, one part corn meal and one part beef scraps. This would need to be supplemented with green food of some kind. Here is another combination: Twelve lbs. corn meal, 6 lbs. wheat bran, 12 lbs. wheat middlings, 10 lbs. meat scraps, 2 lbs. oil meal, 4 lbs. alfalfa meal.

While these various mashes are to be commended to the amateur as well suited to his needs, the fact must not be overlooked that many practical poultry keepers get excellent results when they rely entirely on wheat bran and beef scraps, either mixed in equal parts or fed in separate hoppers, the latter plan being preferred. In spite of what the experts may say about the theoretical value of this combination, there seems to be something about it which makes it exceedingly satisfactory. Many breeders bring up their chicks on it, to a large extent, and all fowls eat it freely.

Patent foods and condiments should be shunned. It seems necessary, though, to make one reservation here. Of late, various experiments with mustard seem to show that it may become an important item in feeding for eggs. Very good reports follow its use in a mash, the amount being about a teaspoonful to the quantity of mash eaten by twenty-five hens in a day. A little salt in the mash is also beneficial. The table scraps may be salted to just about the extent which would make food palatable for human beings.

Only the veriest tyro needs to be told that oyster shells and grit must be kept in hoppers where the hens can have access to them at all times. To be sure, there is considerable disagreement about the necessity of grit when oyster shells are used, one noted expert declaring that months will pass without the hens touching the grit. At any rate, no harm will be done if the fowls are without it for a few weeks, but it is being on the safe side to keep a box in the house at all times. A box of charcoal is also recommended, for charcoal is an excellent absorbent and the poultry seems to keep in better condition when it is always at hand.

An egg is largely water. Without water hens will not lay eggs. There is a string to that statement, too, for they will do very well without water if they have snow to eat. Some poultry keepers warm water for their hens all Winter; others give them no water of any kind when they can get snow. No doubt the hens which have the warm water give their owner a few more eggs than the snow-fed birds, but whether enough more to compensate him for the labor of carrying the water depends upon how valuable his time is.

Some rather foolish statements are made about the necessity of warming the water for hens in Winter. We are told that cold water chills the digestive organs, but when one considers that a hen drinks only about a teaspoonful at a time, one may judge that this chilling is not a very serious matter, after all. It is quite possible that the hen enjoys sipping warm water in the morning just as some human beings do, and everything which tends to make the hen comfortable and happy helps to promote egg production.

Truth to tell, the getting of eggs depends to a very large extent upon keeping the hens in a contented state of mind. Worry a hen or change her surroundings and observe how quickly she will cease to lay. By the same token, then, it is worth while taking care that the flock has cool, fresh water at least twice a day in Summer. It is an advantage to have the water dish in a shaded place outside the house and an iron or earthenware dish will help to keep the water palatable.

The use of deep litter has been mentioned, but the subject is worthy of elaboration. Hens must have exercise and there is no better way of compelling them to take it than to scatter the grain in several inches of straw, shredded corn stalks, hay or leaves on the floor of the poultry house. As the litter becomes packed down, more should be thrown in during the Winter, until it may have become a foot deep by Spring, at which time it should all be removed, and will prove a valuable addition to the garden. Swale hay makes good litter and a bale will answer the amateur a long time. Good straw is considered the best litter, but it is expensive. Leaves will answer and need cost only the few nickels which will reward a small boy for raking them up. Leaves pack hard and need loosening with a rake or stable fork.

Whatever litter is used, the hens have a tendency to scratch it to the rear of the house, and it is a good plan to take five minutes daily in order to scatter it evenly over the floor. Some amateurs rake the grain into the litter every morning.

Poultry are early risers in Summer and need their breakfast as soon as off the perches. If that is too early for their owner, it is a simple matter for him to scatter the grain in the litter the night before.

The importance of green food is often overlooked. It means much in getting a good egg yield and in keeping the hens in condition. There is nothing better than alfalfa, and clover comes next. They may be made a part of the mash or cut into short

lengths and boiling water poured over them, being
served after they have been allowed to steam for
an hour or so.   Indeed, the fowls will eat much dry
alfalfa or clover, and chaff from the barn loft is a
splendid addition to the litter, although still better
for the floor of the brooder or brood coop.   Lawn
clippings are valuable and most amateurs can secure
them easily by attaching a grass catcher to the lawn
mower.   The best way to prepare them for winter
use is to spread them on a strip of burlap or a grain
bag and let them stay in the sun until they crackle
when touched.   Then they may be stored in bar-
rels or bags and will be greatly relished when soaked
or steamed months afterwards, or even when fed
dry.

Lettuce and other greens are available in Summer, but it is a good plan to plant a few cabbages
or mangel wurzel beets for Winter.   The beets are
easier to keep than the cabbage.   Dwarf Essex rape
is excellent and if sown in April will be ready in six
weeks.   If the tops are cut off several times, new
ones will grow, so that this plant offers an easy
means of getting greens in abundance.   Swiss chard
may be used in the same way, renewing itself quickly
when the outside leaves are removed.

The man with but little spare time can make it possible for the hens to pick their own greens in Summer by sowing a patch of oats in the poultry yard and putting over it a strip of inch-mesh poultry netting attached to a light frame resting on two boards six inches wide set on edge at each end, the boards being sunk two inches into the ground to hold them in position. The hens will be able to eat off the oats only as they grow high enough so that they can be reached through the wire.

The plan of hanging cabbages and other vegetables in the hen house so that the hens will have to jump for them is not a good one, although often advocated. The birds are likely to be injured. Hens are fond of kale, which is not injured by the frost and may be left in the garden until after snow flies. This is an excellent green ration for early winter and easy to grow.

CHAPTER V

HATCHING THE CHICKS WITH MACHINE AND HEN

THE way to get good chickens is to begin with the hens that lay the eggs. These hens should be well-matured, hardy, in good condition and come from an egg-laying strain. Not more than a dozen birds of the lighter breed like the Leghorns and six to eight of the heavier breeds should be mated to one male, in order to make sure of fertile eggs. In some cases it is well to have two cocks for each pen, alternating them each week and keeping the rooster not in use in a pen by himself. When this plan is followed the pen may safely contain more hens.

Eggs for hatching should not be kept more than two weeks and they need special care. They ought to be stored where the temperature does not run much below forty or much above sixty-five and where they will not dry out rapidly. Some people make a practice of keeping them in bran; others store them in cake or bread tins with closely fitting covers. Each plan serves to keep the eggs away from the light, which is desirable. It is especially necessary that they be kept in a place which is not at all damp, or they will be likely to be touched with mold. They should not be packed in sawdust. It should not be understood, of course, that eggs kept in a basket on the pantry shelf will not yield any chicks; the point is simply that eggs properly cared for will give a higher percentage of strong youngsters than those which are neglected.

It is rather better not to set white and brown shelled eggs together, as the latter have slightly thicker shells, which fact may serve to prolong the hatch. Eggs of uniform size, smooth and without any abnormal features are the ones to select.

Probably the average amateur will use hens for some years to come. The incubator is a wonderful invention and indispensable on large plants, but the man hatching from fifty to one hundred chicks will doubtless find it an advantage to rely on hens. When a larger number of chickens is desired or when they are wanted very early in the season, the question of investing in an incubator should be seriously considered.

Some amateurs keeping hens of the non-sitting breeds may, indeed, prefer to use a 70 or 120-egg machine to hatch their chicks rather than to bother with broody hens obtained from the neighbors. These little machines are thoroughly practicable and not difficult to operate.

It is sometimes a desirable plan to set several hens at the time that an incubator is started. Then, at the end of ten days or two weeks, the eggs may be taken from the hens and placed in the machine to take the places of the infertile eggs tested out. Of course the eggs under the hens should also be first tested, and in this way the twenty-first day brings a machine full of chicks in no danger of being stepped on by a blundering hen, and free from lice. It is also found entirely feasible to hatch the chicks in an incubator and raise them under hens just as it is to hatch them under hens and raise them in a brooder. Both plans are practiced.

Hatching chicks with hens has one distinct advantage. With 70 fertile eggs set under five hens, one is almost certain to get a fair number of chicks, for even if one hen abandons her nest and another crushes her eggs by her clumsy movements, there will still be an excellent chance of obtaining a satisfactory number of chickens from under the three hens remaining, while if 70 eggs are put into a machine and anything goes wrong, the loss probably will be total. Such loss becomes greater in proportion to the increase in the size of incubator used.

Setting a hen should be a matter of some care, but it need not be the solemn rite some people make it. A common and simple plan is to arrange a row of commodious boxes in a quiet place and make the nests for the sitting hens in them. The boxes may be set upright and in a row with one board over all of them to confine the hens, or they may be set on their side with a board in front. The use of a single board makes lighter work than the construction of a door for each box. The board may be removed at a certain hour each day and the hens allowed to eat and drink and to dust themselves in the box of earth or ashes which should be provided for them.

Some breeders have a little pen in front of each nest and allow the hens to come off when they please, which means the saving of a little time. Sometimes, though, a hen will not voluntarily leave the nest as often as she should, in which case she must be gently ousted. If the hen does not eat she will become greatly reduced in flesh.

When a hen becomes broody and the owner desires to set her, she should be moved from the laying house at night and placed in the nest prepared for her. A nest egg may be placed under her, and her actions when morning comes will determine whether she shall be trusted with the eggs to be incubated. If she is found sitting tight and manifesting the customary signs of anger when disturbed, she may be given the eggs. It is always best to put the eggs under the hen rather than to put the hen on the eggs.

The nest itself should have a shovelful of earth at the bottom, if possible, with a generous supply of fine hay above. It should not be concave, for then the eggs will roll to the middle and not separate easily if the hen attempts to put her foot between them. If the nest is made almost flat with a ridge at the outside to keep the eggs from rolling out, there will be less danger of broken eggs and yet the hen will get all of them under her.

A free use of some good insect powder like Persian Insect or Dalmatian powder, which may be bought of any druggist, should be made when the hen begins to sit and once a week thereafter.   Hens often die when sitting because of the inroads made upon their vitality by lice.   It is cruel torture that a lice-infested hen endures and the owner should make it unnecessary by dusting her thoroughly, particularly under the wings and around the vent, which work may easily be done if the hen is held by the legs, head down.

A plan which makes feasible the setting of hens in the laying house has some things to commend it.   Two rows of nests are made, one above the other, with a hinged board so arranged that it may be used to stop the entrances to the top tier or to cover the openings leading to those below.   Sitting hens are placed in the top nests and the board raised so that they are confined.   The board also prevents other hens from getting into the nests, either to lay or to annoy the hens which are sitting.   At the time of the afternoon feeding, the attendant turns the board down so that it covers the lower tier of nests. Then the sitting hens are free to fly down and feed with the other birds.   When they are ready to return, they are obliged to enter the top nests because the lower ones are protected by the board.   If as many nests are vacant as there are broody hens, each hen will soon be settled again on a setting of eggs, although perhaps not in the nest which she left, for hens seem to have a very short memory; it is not uncommon for one to leave a nestful of hatchable eggs and sit on a China egg or two. The advantage of this method lies in the fact that the amount of work required is minimized, while hens are much more likely to be satisfied when changed from one nest to another in the same house than when moved to strange quarters. One can never be sure that a hen moved from one locality to another will continue to sit.

When a man runs an incubator, he puts all his eggs in one nest, as it were. Then he has one machine instead of a number of hens to look after. Very little work is required, and that not of an arduous nature, but painstaking attention to details is imperative. Sitting hens will tolerate a certain amount of neglect because they are able to adjust themselves in some degree to circumstances. When using a machine, however, all the intelligence must be manifested by the operator.

It is not wise to buy any but a standard machine — such a machine as is generally used on large plants, which can afford to test the different

makes.   It may hold from fifty to about 300 eggs.
Generally speaking, it is advisable to use an incubator holding at least 120 eggs, for it will require
no more attention than a smaller one.   It may be
operated in a cellar, a room in the house or an out-
building.   A fairly even temperature and no drafts
are to be desired.   The machine should not stand
close to a window or where it will receive direct
sunlight, for those reasons.   The cellar of the
house often makes an ideal location, but it should
be well ventilated.   Probably the lack of fresh air
in abundance is one of the most common causes of
poor hatches.

If the amateur decides to purchase an incubator
and operate it in his home, it is well for him first
to consult his insurance agent; otherwise, he may
have serious difficulty in collecting his insurance
money in case of fire from any cause.   It is true
that incubators sometimes get afire, although almost always for the reason that they have not been
properly cared for, and insurance companies exact a
small fee for the privilege of using them.

Before the eggs are put into the machine it should
be run for a day or two, so that it will become
thoroughly heated, as well as to allow the operator

to become better acquainted with the simple mechanism. It should be regulated so that the thermometer will show exactly 103. And it is an excellent plan, by the way, to have the thermometer tested by a physician or druggist, for it is not unusual to find one which is not just true and if the amount of variation is known, it can easily be allowed for. Strangely enough, the fault, if any exists, is generally found at the point between 100 and 103 degrees. Faulty thermometers cause frequent losses.

The kerosene oil should be of the best quality that can be obtained. Whether to fill the lamps at night or in the morning is an open question. If filled in the evening one is sure that a strong heat will be carried through the night, but if filled in the morning, it is easier to check any tendency of the flame to run up. A new wick should be used at the beginning of each hatch and the lamps should always be kept perfectly clean, with the wicks trimmed daily. Too careful attention cannot be given.

After the second day, the eggs should be turned night and morning. This does not mean that they must be directly reversed, but that they should be shifted about in order to ensure an even distribution of heat. Some people tack a piece of cardboard marked "Day" to one side of the tray and another marked "Night" to the other side and make it a point to have the former show when the eggs are put back in the morning and the latter when they are returned to the machine in the evening.

When the eggs are turned, they may also be aired or cooled. This is an important matter, for it helps to develop strong chicks. While the eggs are out of the machine, the door should be kept closed. The extent to which the eggs should be cooled depends upon conditions. Of course, they can be left out only a short time in very cold weather. Perhaps a safe plan for the amateur is to place a thermometer on the eggs as soon as they have been turned and restore them to the machine when the mercury has dropped to 85 degrees. It is not necessary to become unduly alarmed if the eggs are permitted through inadvertence to become cold. The writer once forgot a tray of eggs until they had been out an hour. Wondering if it would be of any use to continue the hatch, he broke an egg and found a live chicken. The eggs were returned to the machine and heated up quickly, with the result that an average hatch was secured.

When the eighteenth day closes, cooling and turning of the eggs should cease, for the chicks are almost ready to hatch. In very dry climates, the eggs may then be sprinkled with water at a temperature of 103. Much might be written about the moisture question, but experts differ and the safest plan is to follow implicitly the directions which come with the machine.

On the seventh day the eggs may be tested, and it is well to perform this operation again on the fifteenth day. The test is made by placing the eggs between the eye and a strong light and excluding all other light. In practice, a tester which fits over a lamp chimney and allows light to come only through an opening at one side is the easiest to handle, as both hands are left free. When an egg is placed tightly against the frame around the opening, the contents will be illuminated. If the egg is infertile, it will be entirely clear; if there is a chick in it, an opaque spot will indicate its presence. On the seventh day this spot will be small and lines will radiate in all directions. These lines are blood vessels running out from the heart. On the fifteenth day the chick will be large enough to almost fill the shell, appearing as a dark mass. An egg containing a dead germ will be known because it is not clear like an infertile one, and yet has no blood lines when the seventh day test is made. These eggs should be removed and thrown away. The clear eggs may be saved and boiled hard for the newly hatched chicks. If a considerable number of eggs are tested out, eggs from under hens set at the time the machine was started, may be substituted. It is always well to test the eggs under sitting hens in the same way. Then, if the fertile eggs are not desired to replenish a machine, it may be possible to give the eggs from two hens to one, so releasing one hen.

A workable tester may be made at home with the aid of a breakfast food box. The top should be removed and the sides cut away so that the end may be made to closely cover the face, shutting out all light. Then a hole slightly smaller than an egg may be made at the opposite end. If this little device is held toward a strong light and an egg placed closely against the opening, the testing can be done very quickly and easily.

Another way of testing eggs where there are a considerable number is recommended. A light board is fitted over a window facing the south and a hole made in the board. Then, when the sun is shining brightly, it is only necessary to darken the room and hold the eggs over this hole in order to test them. If a black curtain fastened to the board is dropped over the head, it is not even necessary to have the room dark.

Eggs are often pipped on the twentieth day and the chicks should be out by the end of the twenty-first day. However, a hatch is delayed or prolonged if the temperature has been low, just as it may be hastened by running the temperature high. It is a normal hatch when all the chicks appear within a few hours and at the proper time.

It is well to restrain one's impatience and keep the door of the incubator closed while the hatch is going on, except that near the end some of the shells may be removed. The natural heat of the chicks is likely to send the thermometer up near the close of the hatch and the temperature must be regulated accordingly. It is not wise to remove the chicks until they are well dried, for they will not need food for several hours and are better off in the machine, if the latter is not allowed to get too hot. Hatching a lot of chickens in an incubator is always an interesting experience, for the whole process is one of the most wonderful of Nature's mysteries.

CHAPTER VI

BROODING THE NEWLY HATCHED CHICKS

MUCH that is written about the handling of newly hatched chickens has little or no significance for the average amateur, because he will not have early broods to deal with. Chicks coming into the world in January and February appear at quite an unseasonable date, according to the laws laid down by Dame Nature, and an exceptional amount of time and care must be given them. These are special purpose chicks, intended for broilers or roasters, and are grown by commercial poultrymen with the proper equipment for raising young stock when the ground is covered with snow.

The amateur poultry keeper, on the contrary, will delay hatching until March or later, and will find less difficulty in raising his chicks. This is especially true if the weather is such that he can get them onto the ground at any early age. Hundreds of chickens are killed by kindness. Too much coddling is as fatal as neglect.

Chicks running with a hen occasion but little trouble. The coop should be large enough so that the hen will have plenty of room to move about without trampling on the chickens, and should have a board floor if out of doors early in the season. After the weather becomes warm, earth floors are better. Sand should be spread over the floor and over that, after the first few days, a litter of hay cut into short lengths or chaff from the barn, the latter being preferable to anything else.

The chicks will need nothing to eat for at least thirty-six hours and may go longer without suffering at all with hunger. Some kind-hearted but mistaken people have insisted that it is cruel to withhold food from a newly hatched chick, but experience has shown the wisdom of doing so. It is true that a chick will try to eat as soon as it can balance itself on its legs, but this is not because of a sense of hunger, and it is just as well satisfied with sand. That is the reason the floor of the coop should be sprinkled with sharp sand; the chick eats a little and so obtains the grit which will aid it in digesting real food when it comes.

Truth to tell, chickens seem practically devoid of intelligence the first week of their existence. They will peck at anything within reach, having an especial fondness, apparently, for the eyes of the hen mothering them, and will eat sawdust as readily as anything else. At the end of seven or eight days they appear to reach the age of discretion, for after that time they cannot easily be fooled on the subject of food.

Some persons make the feeding of young chicks a highly complex matter. It need not be unless they are to be raised for a special purpose, aside from that of producing eggs. Hard-boiled eggs are the time-honored first meal, after the sand, and make a ration which is entirely satisfactory, but not necessary by any means. If the infertile eggs tested out have been saved, it is wise and economical to use them in this way. Bread soaked in milk and squeezed fairly dry may well be given the first two or three days. One of the best plans is to feed very young grass or, better still, clover cut into very short pieces. Too much stress cannot be laid on the value of grass and clover for young stock of all ages. Common oat meal is an excellent ration for chicks just hatched and may be fed freely.

After two or three days of using baby feeds of the kinds mentioned, it is customary to change to regular chick grain. As a matter of fact, the commercial chick foods, which are altogether the best for the amateur to use, may be fed from the very first. After a few weeks he can change to cracked wheat, cracked corn and Kaffir corn if he desires, or he may keep right on with the commercial chick feeds, which are a combination of many grains, including kinds which the chicks like particularly well and which they will work hard to get. If a soft feed which is likely to sour is used at first, feeding time should come four or five times a day, but if oat meal or the regular chick feeds are depended upon, three times a day is often enough from the first, when the chicks are with a hen. There may be grain in the litter all the time, but no harm will be done, for the chicks will be guided by the hen to a large extent in the matter of eating and she seems to have a proper instinct about these matters. If there is too much food in sight, she probably will cover it up.

After the tenth day a hopper of ground feed and beef scraps may well be kept within reach of the chicks at all times. They will eat a lot of it and

will thrive on it. This feed may be one of the prepared growing rations, so called, sold at the stores
and containing many kinds of ground grain, as well
as beef or fish scraps and alfalfa, or it may be nothing more than plain wheat bran with ten per cent.
of good beef scraps added. Many good chicken
growers are well satisfied to use this very inexpensive ration year after year.

The value of grass has been mentioned. Green
stuff of some kind is essential. Broken pieces of
lettuce and other vegetables are relished. Cut alfalfa may be bought if there is nothing at home
available, but a little cold frame will serve to grow
plenty of lettuce early in the season. Sprouted oats
are excellent. They are prepared by soaking the
oats over night in warm water and then spreading
them in a box, having holes for drainage, so that
they will cover the bottom one or two inches, and
keeping them moist by sprinkling them daily. If
kept in a warm place, they will soon throw out
sprouts and may be fed when the sprouts are an inch
long. It is better not to feed the roots to very young
chicks.

Water should be given the chickens from the first
day, but in a receptacle of such shape that they can-

not climb into it and get wet. Chick fountains may be purchased cheaply, but are easily made at home by the combination of a tin can and a flower pot saucer. One end of the can should be removed and one side at that end pushed or hammered in. Then the can may be filled with water, the saucer placed over it and the can quickly inverted. The saucer will be found partly filled with water and more will run out as that is consumed. The same result is secured by making a hole in the side of the can just below the edge of the saucer, the latter being large enough so that the chicks can drink around the sides of the saucer. Many people adopt the simple expedient of filling a small dish with water and putting a half brick or a stone in the middle, so that the chickens cannot get into the water.

Chick grit should be kept in the coop and it is also well to have a little charcoal where it is always accessible. Nothing has been said about wet mashes for chicks. All that need be said now is that they are not to be recommended, especially for the amateur, who wants the nearest approach to a safe and sure method of raising his chicks. Yes, it is true that our grandmothers fed sloppy mashes and raised a fair percentage of their chicks. It is a mystery, though, how they did it. And anyway, what is the use of going to the time and trouble of mixing a mash when a little dry feed can be scattered in the litter in a quarter of the time?

It is usually considered wise to keep the hen confined to the coop while the chicks are allowed to run at large, but the youngsters should not be allowed their freedom until the dew has disappeared from the grass, if they have a grass run. And the hen — patient old mother — should not be forgotten. Plenty of whole or cracked corn and some oats may be given her, at first in a tall dish that the chicks cannot get into, and she should have water always at hand. Likewise, she should be dusted at least once a week with a lice powder, well worked into the feathers, so that the chicks will get some benefit from it.

Chicks with a hen are certain to suffer from the plague of lice. If neglected, they may die from the inroads upon their vitality made by these pests. When chicks stand around moping, it is time to get out the dusting box. The chicks ought never to reach this stage, though, for an application of powder made at night when the hen has been lifted off, should be a weekly practice, beginning the first day.

A very little lard rubbed on the head of the chicks will help, too.

Chicks reared in a brooder require more attention than those raised with a hen, although, happily, the lice nuisance is escaped, at least, for the first few weeks. The lice seem to appear even on incubator and brooder chicks after a time, coming from nobody knows where. However, when a hen has the chicks, you can scatter chick feed generously in the litter and go away for the day with a reasonable expectation that biddy will look after matters while you are gone and that the youngsters will be safe and happy when you get back. But you can't play truant when using a brooder. Feeding must be done regularly and the heat must be properly regulated. If the chicks get too hot or get chilly, the results are likely to be equally unpleasant. And yet, running a brooder is not an irksome or difficult task. Of course, something depends upon the time of year. If the season is well advanced and the chicks can be put on the ground early, the work is made easier. There are both indoor and outdoor brooders. The latter may actually be used out of doors with the ground covered with snow and in zero weather, but caring for them is not a job to be welcomed. Later in the season, they give excellent satisfaction.

The chicks should not be removed to the brooder until thoroughly dry, and it is well to start the brooder lamp by the time the eggs begin to pip, so that it will be nicely heated and ready for the chicks as soon as the chicks are ready for it. The floor should be sanded like the hen's coop and a supply of clover or alfalfa cut into short lengths for litter provided. The heat should be from ninety-five to a hundred degrees for the first week and decreased at the rate of five degrees a week thereafter, making the reductions gradually, of course. When the chicks are put into the brooder they will run up the temperature several degrees, which should be anticipated.

Although the thermometer is necessary, observation will determine more accurately the degree of comfort which the chicks are enjoying. If they are found stretched on the floor and panting, the heat is too great; if they huddle closely, it is insufficient. If they settle down contentedly slightly apart, perhaps with some heads sticking through the felt, they are satisfactory proof that the heat is just right. Lack of ventilation is a frequent cause of trouble. The brooder chicks need fresh air in abundance. This point is too often overlooked.

After a few days the chicks may be allowed brief excursions outside the hover, but in order to prevent their getting lost, it is well to make a little yard of poultry netting arranged in a half circle, so that as a chick moves along it, he will be guided back into the hover and comfort. Corners should be avoided; chicks get into them and huddle there until chilled.

The same kind of feed as that described for broods with hens will answer as well for brooder orphans. It should be fed regularly, though, and four times a day for two weeks. Too much should not be given at a time, the purpose being to keep the youngsters busy from daylight until darkness, except at napping times. They will soon learn to scratch in the litter. The mash should be given at ten days. If a commercial growing mash is not available and something more than bran and beef scraps is desired, a very good mash may be made by combining three parts of bran, one part of corn meal, one part of middlings and one part of high-grade beef scraps. There is an easy way of mixing a mash of this kind, which is worth knowing about if there is much mash to be made. If a revolving churn is secured and the various ingredients put into it, a few brisk revolutions of the handle will blend the mixture perfectly.

Fireless brooders are much in evidence and apparently have come to stay. Although not having the wide latitude of usefulness first promised, they often can be made to serve the amateur very well indeed, and cost only a dollar or two. Indeed, such a brooder is easily made at home with the aid of a cheese box, from which top and bottom have been removed. An opening is cut for the chicks and a piece of burlap tacked over the top, being allowed to sag in the middle. The interior is partly filled with hay and as much padding placed on top as may seem necessary.

A well-ventilated room is the best place for the fireless brooder until the weather becomes warm, when the porch proves an excellent location. Fresh air in abundance is most important. The chief difficulty in the use of fireless brooders is to induce the chicks to go in after they all have come out. Of course, the brooder is warm only when it contains the chicks, for it is the plan of the device that the occupants shall generate their own heat. Consequently, when it is not warm, there is no inducement for the chicks to enter. This difficulty is overcome by placing a hot water bottle on top of the brooder. That provides the warmth needed and acts like a magnet. Once the chicks get the habit of running into the brooder when cold, they will keep it up after the water bag has been removed and will keep each other warm. When only a few chicks are to be raised and after the season is well advanced, the fireless brooder is to be recommended.

After the chicks are gotten onto the ground, they will make rapid progress, but must be protected from hawks and cats. There is no better place for them to run than a patch of corn, where they will be safe from the hawks and will have adequate shade. Wherever confined, shade of some kind must be given or growth will be checked. An orchard run is good, but it must be remembered that after grass gets old, it is so tough that young chicks cannot eat it, and so must be given other green stuff. When the young birds are eight or nine weeks old, the sexes should be separated for best results. The surplus cockerels should be fattened and marketed, and by the first of October the pullets should be in their winter quarters.

CHAPTER VII

HOW TO GET EGGS IN WINTER

WHEN eggs are sixty cents a dozen, the amateur finds no little delight in bringing in a daily basketful from his poultry house. Unfortunately, though, he is often denied this privilege. All too frequently the hens refuse to do their part.

Getting eggs in Winter is a problem which has received much attention. It is no longer a matter of hit or miss. Given pullets which reach the laying age before settled cold weather, house them in their permanent winter quarters by the first of October, keep these quarters dry and free from drafts, but with fresh air entering in abundance at all times, keep a deep litter on the floor so that the hens will be obliged to scratch energetically and persistently for their grain, give them a wide variety of rations, and the eggs will be reasonably sure to come. The better the strain, the larger the number of eggs.

84

There are no secrets about the production of winter eggs. It is just a matter of foresight and care. Pullets undoubtedly are the best layers. Hens in their second year will lay fairly well, but it is better to start fresh with pullets each season. Hens that molt very late will not prove profitable to keep. If pullets are hatched too early, they may molt the same season, which is not desirable. And yet if hatched too late they will not begin laying until after the New Year. Leghorns will lay when from five to six months old. Some Plymouth Rocks commence laying at six months. The other hens in the same classes begin laying at about the same ages. The larger breeds like the Brahmas require more time before producing their first eggs. In order to make sure of winter eggs, the pullets should reach the laying age in October. If they are neglected during the Summer, they will not lay as early as though given good care, which is entirely logical.

As a rule, hens lay better in flocks of not over thirty. And yet, some of the most successful egg farmers keep 500 birds together and get satisfactory results. Long houses give the fowls plenty of room when there are no partitions, which is an advantage. The average amateur, however, will not carry more than fifty hens, so that this matter of large flocks will interest him only in an academic way.

The use of litter on the floor is most important. It may be two or three inches deep at the beginning of the season and more thrown in as the first becomes broken into fine pieces. Of course, there is such a thing as having it too deep. The main thing is to keep the hens working early and late seeking food. A little millet or hemp seed in the litter will act as an extra inducement to scratch energetically and persistently.

It must be remembered that in mid-Winter the hens keep short hours. With two-thirds of their time spent on the roost, they should have no time to waste during the day. The poultry keeper wants them to eat all they will and a busy hen has a much better appetite than one which stands around idly.

Green food is essential. Sprouted oats have been mentioned in another chapter. Wheat and barley may be sprouted in the same way. Some amateurs who are handy with tools make a little frame which contains four or five trays on which the grains are spread after they have been soaked over night in warm water. The bottoms of the trays have slats placed so closely together that the grain will not pass through although the water will drain away, or are covered with copper screen cloth, copper being used because it does not rust. The water sprinkled on the top tray will work down and wet the grain in all the trays underneath.

Of course the hens should be given all the water they need, as well as grit and oyster shells. A box of charcoal is also worth while. It is even more important in Winter than in Summer to provide everything that is needed for the making of eggs and to keep the hens in first-class condition.

Few people realize to what extent eggs are affected by the food given. An experiment was tried at one of the agricultural colleges a few years ago. Limburger cheese was fed to a number of hens and when the eggs from these hens were broken, the odor alone was sufficient to prove the truth of the theory. Some of the eggs strayed to the president's table, it is reported, through an oversight, and — well, what the president said has not been recorded.

In order to have the highest grade eggs, and of course the amateur wants no other kind, only the best grain should be fed. Likewise, the water should be fresh at least once a day. It is well to gather the eggs twice a day, at least; otherwise some of them may be incubated by the succession of laying hens for several hours. If the eggs are fertile, this is enough to start the germ into development.

As a matter of fact, it is better to have no male bird with the flock. There probably will be more eggs and less danger of broken ones. If an egg is broken in the nest, it should be removed from the pen. If thrown on the floor, the habit of eating eggs may be established. The nest should be thoroughly cleaned out and fresh hay substituted.

The color of the yolks is influenced by the feeding. Yolks which are yellow to an objectional degree indicate an almost exclusive diet of corn. Alfalfa, clover and grass clippings tend to give the yolks a rich shade of yellow, just as they affect the color of milk. Rape fed to excess gives a very pale shade to the yolks, over-much cabbage makes them thin.

Eggs should be kept in a cool, dry place. Dryness is very important because germs cannot penetrate the membrane of the egg unless it becomes moist. Sometimes eggs absorb the odors of other food stored close by, if highly scented like bananas and onions.

It is useless to expect the hens to lay well if they are preyed upon by lice, so that a dust bath is very necessary, unless there is an earth floor into which the birds can burrow. A few upright boards may be fastened together to make a dusting place and earth or ashes with a little lime added given for the dusting material. Coal ashes are good except that they tend to rob the plumage of its luster. Hens like coal ashes and will eat many of them. Dry sand is often used, but the fowls seem to prefer heavier earth. It is well for the amateur to lay in a barrel of earth or road dust in the Fall to be used in the course of the Winter. Whenever hens are purchased, it is advisable to give them a thorough application of lice powder, dusting it with a generous hand into the fluff around the vent.

If the fowls are slow in beginning to lay, a little green cut bone may help start them. It is doubtful whether the average amateur is justified in the purchase of a bone cutter, but in many large cities it is possible to buy green cut bone ready for use. Of course, it must be fed at once, as it will keep but a short time.

Another plan is to try feeding a warm crumbly mash, containing a liberal amount of beef scraps or

meat of any kind, once a day two or three times a
week. A teaspoonful of mustard for each twenty-
five hens may be included in this mash, which, for
the rest, may be made of two parts bran, one part of
ground oats and one part of corn meal.

A laying hen is usually a singing hen. Likewise,
the hens which are off the roost first in the morning
and on the last at night may be put down pretty
safely as being good layers. It pays the amateur
to spend a little time with his flock; he can learn
a lot in that way. When a hen is laying well, her
comb is full and bright red. She may begin laying before her comb gets its color after the molting
period, however, but it will gradually become fiery.

There should be enough nests so that the hens
will not break the eggs by crowding. It is a general rule to allow one nest to five hens. If the
amateur seeks to build up an egg-laying strain, he
can make use of trap nests providing he has sufficient time so that he can give them the attention
they require. These nests hold the hens which have
laid until an attendant has released them and by
banding a leg on each hen and keeping a record, it
is possible to tell just how many eggs each hen lays

in a given period.   Then the best egg producers
may be used for breeding stock.

With a trap nest it is possible to identify the hens
which lay the large eggs and those which go onto
the nest but seldom lay an egg.   In fact, it gives
the poultry keeper a working knowledge of his flock
not to be obtained in any other way.   Trap nests
require close attention, of course, but not so much
of the poultry keeper's time as might be imagined,
if there are nests enough.   Allowing a nest to every
four hens, he will not have to visit the poultry
house oftener than four times a day, and no hen will
be confined long enough to suffer, unless the weather
be very warm, and trap nests are not commonly
used in the warm season.

There is a simple plan by which the results secured by the trap nest may be approximated with
but little trouble to the attendant.   A box containing a nest is placed in the partition between two pens,
one end of the box being open while a trap door is
arranged at the other end, so that the hens can enter but not leave that way.   The hens may be put
into the first pen in the morning and as fast as
they lay will pass into the second pen, so that when

night comes it is an easy matter to decide which hens have laid during the day. If a male bird is placed in the second pen, the hens which lay and pass into that pen may confidently be used as breeders. Of course, some hens will go onto the nest but not lay, so that this test is not quite as accurate as that imposed by a regulation trap nest.

The average amateur, however, will hardly take the trouble to trap-nest his birds. A simpler plan is to select and mark the pullets which lay first in the Fall and use them to breed from. Experiments have shown that the pullets which begin laying earliest also make the heaviest layers, as a general rule. These pullets may well be kept until the second year and then mated with a well-developed cockerel.

Poor flocks may be improved by securing a male bird from a breeder who is known to have a good laying strain. Yet, it is not wise to continually introduce new blood. If the first cock bought for improving the flock proves satisfactory and another is needed later, it is well to secure it from the same source. Males from eggs laid by heavy-laying hens are to be sought. They transmit the trait to the pullets they sire.

Old hens should not be kept with pullets as a

rule.   They require rather different treatment.   If you are going to keep over a number of extra good hens to use as breeders in the Spring, they should not be forced as hard as the pullets.   Fowls forced for eggs are not in proper condition to make good breeding stock when the breeding season comes. All the other left-over stock should be disposed of before the pullets are put into winter quarters. They will bring low prices if kept until November or December.

It will be seen from all this that there is no royal road to winter egg production.   It is all a question of properly hatched hens, properly reared and properly fed.   The requisites are not numerous, after all, but they are exceedingly important.

CHAPTER VIII

KEEPING POULTRY ON A TOWN LOT

THE no-yard system makes it possible to keep a few hens on a very small lot. To be sure, this is not the best system, all things considered, but it offers a very satisfactory solution of the poultry-keeping problem where only a little land is available. By a little land is meant enough to provide the site for a house 10x12 or smaller, with sufficient open space around to admit air in abundance and sunlight for at least four or five hours every day. When this system is followed, the hens never leave the house. In order that the birds may be kept in good condition when confined so closely, the house should have unusually large openings in front, with muslin curtains to drop when the weather becomes exceptionally cold or when the rain beats in. If the Summers are very hot, something in the way of an awning may be needed to protect the fowls; a hinged shutter is often used. The house may also be made cooler by cut-

94

ting an opening in the rear wall just under the roof, with a shutter to cover it when the weather is cold.

At least four square feet of floor space should be allowed each bird when this system is followed, and the nests and all the furnishings should be high enough so that the hens can walk under them, making the whole floor area available. It is customary for amateurs who keep hens on the no-yard plan to buy pullets in the Fall, and to dispose of their old hens as fast as they stop laying in the course of the Summer. Before the new flock is installed, the house should be thoroughly sprayed with a lice paint or with kerosene in which a little carbolic acid has been mixed, or the interior may be whitewashed.

New litter should be substituted for the old, and it is well to replace an inch or two of the earth, if earth floors are used, with fresh sand. Cleanliness is one of the most important matters when hens are confined closely and the amateur will inevitably find that eternal vigilance in this matter is the price of success, especially in Summer, when vermin multiply with exceeding rapidity. It will be necessary to remove all the fixtures frequently and wash or spray them with kerosene or a liquid lice killer and to frequently renew the nesting material.

A spray pump is a great convenience as well as a saver of time. An air sprayer, which may be purchased for less than eight dollars, is especially desirable, for it may be charged with a few strokes of the plunger and then slung over the shoulder by a strap, while the operator guides the stream in any desired direction and regulates it with a thumb screw. Whitewash may be used in this machine, if it is mixed thin, and the amount of time needed to cover the walls greatly reduced. Incidentally, a spray pump may also be used to advantage in the garden when insect pests make their appearance.

When the no-yard system is followed, the floor must be kept covered with litter at all times, for the hens must be induced to exercise. And of course there must be water always at hand, for each laying hen averages to drink half a pint a day when the weather is warm. The water dish ought to be refilled several times a day in Summer, if this is feasible, in order that the hens may find the water palatable. In Winter, the water is likely to freeze after a short exposure to the cold. Water as warm as the hand can be borne in it may be given, in order to lengthen the time which will elapse before it turns to ice.

Often the problem is not lack of land as much as lack of time. Many commuters would like to keep a few hens if they did not find it necessary to leave home early in the morning, perhaps with no assurance of getting home again until after dark. If the wife or some other member of the family may be interested in the hen-keeping project, the birds will not suffer for lack of feed and water; but it frequently happens that nobody in the family wants to bother with them — and the work is a bother unless one has a genuine liking for well-bred poultry.

There is a way of meeting this difficulty and keeping even a good-sized flock with only ten or fifteen minutes attention each day and with an extra hour on Saturday or Sunday, when a general cleaning may be indulged in. By means of a patent feeder and exerciser which costs $2.50 and a patent water fountain costing one dollar, combined with the use of hoppers for dry mash, as already described, the commuting poultry keeper can entirely dispense with daily feeding and watering.

The feeder holds from eight to thirty-two quarts of grain, a few kernels of which drop out every time a bait bar under the machine is moved. This bar is made of wire netting and filled with cracked corn. None of this corn escapes, but the hens see it and peck at it. The slight blow is enough to turn the bar a trifle and down comes a shower of corn, which is scattered by a deflector in a wide, even circle. If there is a little litter on the floor, the hens will scratch in lively fashion for a few minutes, after which another peck at the bait bar will result in another deluge of grain. In this way, the fowls are kept active and there is no waste of feed. All that falls out is eaten and neither mice nor birds can extract any from the feeder. When the hens tire a bit of this exercise or of the food which it brings them, they turn to the hoppers of mash. These hoppers may be large enough so that they will not require filling oftener than once a fortnight and the grain feeder will contain enough for from two to three days to a week or more, depending upon the size and the number of hens using it.

The water fountain is attached to a butter tub, which is the reservoir, one filling of which will suffice for several days. The tub is covered, so that the water is kept clean, and the fountain is so constructed that the water is always several degrees cooler than the atmosphere. This is, of course, a summer arrangement. For Winter, a fountain with a safety lamp attachment may be used, the heat being just sufficient to keep the water from freezing at any time.

Such devices as these simplify poultry keeping to a remarkable extent, and eliminate the " haven't time to look after hens " excuse. And the flock does not suffer when they are used, the disadvantages being the fact that the hens do not become as tame as when an attendant is among them frequently and that it is not so easy for the owner to observe the condition of the various individuals in the flock.

Feeder and fountain may be used out of doors in Summer, if deemed desirable, care being taken to place the fountain in a shaded spot. The feeder gives excellent results when standing on sod, the grass taking the place of litter. All of these devices may be used for chickens as well as for mature hens, although it is hardly a wise plan to let the young stock go all day without being looked at occasionally, to make sure all things are going well.

Plowing or spading the poultry yard is the ounce of prevention worth a pound of cure. Probably no one cause has resulted in the untimely demise of more chickens than tainted ground. If possible, the youngsters should be raised each year on ground which has not been used for poultry since a green crop of some kind has been grown on it. Plowing up the land and sowing oats or rye will help to purify the soil. If the land used for the young chickens can be planted to winter rye in the Fall, a double purpose will be accomplished, for the ground will be put into condition for chicken raising the next season and the rye will give the hens green food in Winter, for they can be allowed to range over it when there is no snow on the ground.

Where there is little land, the chicks with hens may be confined in small coops easily made of dry goods boxes with a chicken wire run. The end of the run may be divided into a small feeding compartment for the chicks by making a partition of laths far enough apart so that the young birds can pass freely through. If fed here the chicks can be given any kind of food and it will not be wasted or spoiled by the hen. If coop and run are made solid, that arrangement is an advantage, for the whole outfit can easily be moved by two people and a shift every few days will keep the chicks always on fresh ground. Even when it is safe to give the chicks free range, it is best to confine the hen if the poultry plant is on a town lot.

Sometimes it is difficult to get enough green stuff for the hens and chicks if the town lot is a small one. Usually, though, it is possible to find a few feet of ground where Dwarf Essex rape can be sown. Seed put into the ground on Decoration Day has yielded a cutting by the first of July, an illustration of the rapidity with which this crop grows. A few rows of rape will produce enough green food to supply a small flock of hens all Summer.

In the Fall, it often is possible to buy imperfect heads of cabbage for which farmers or market gardeners have no regular market, and at a very low price. The cabbage may be stored on the north side of a building under a foot of soil, with straw, leaves or cornstalks as additional protection.

A considerable number of mangel wurzel beets can be raised in a small space and may be stored in any vegetable cellar. It is not economy to cut them into small pieces; a better plan is to split them in half and drive a spike through them into a board. Then the soft part will be eaten out without loss. A hot bed is a decided advantage, if the man with a little land has time to look after it, for lettuce can be grown all Winter. Swiss chard started in the Summer can be kept along several months by covering it with a cold frame.

If the town-lot poultry keeper cares for the comfort and craves the respect of his neighbors, he will make it a point to keep his hens and chickens confined to his own premises, and he will not have a rooster. The matter of fencing is important, for some hens fly high. Yet a very high fence is objectionable. If six feet of wire will not keep the fowls out of the neighbors' garden patches, a strip of netting a foot wide should be run around the top, covering the yard to that extent. When the hens try to fly out, they will meet this obstruction and be thrown back. Most hens find it difficult to scale a fence unless they can see the top and so gauge their distance. For that reason, there should never be a bar at the top of the wire. If a bar is needed for appearance or support, let it be more than a foot below the top.

It is a great convenience to have a gate wide enough so that a wheel-barrow will pass through and to have it swing both ways, with springs to shut it. When an amateur puts up a poultry fence, the gate usually gives the most trouble, if he does the work himself, so it is well to know that a gate without the wire may be bought for one dollar.

When a single hen escapes from a poultry yard, she commonly displays as much anxiety about getting in again as she did about getting out. Yet she is not willingly cornered and caught. There is a way to get such hens back into the yard without any effort on the part of the owner. As all poultry keepers have observed, a hen will run along the entire length of a wire fence, pressing against it and trying to find an opening. Let the amateur poultryman make a little gate and fit it over an opening at the bottom of the fence just large enough to admit a hen. Let him have this gate open into the yard only, and so hung that it will close automatically but yield to gentle pressure. This may be done by the proper placing of the hinges. As the hen outside the pale pecks along the fence, she presently comes to this little gate. Finding that it yields, she pushes against it a little harder. Behold, it flies open. She walks in and the gate closes behind her. Obviously, it cannot be opened from within and so makes a perfect self-acting trap for wandering birds.

A flock of twenty-five hens given intensive culture, as it were, on a town lot, should produce a considerable larger number of eggs than the average family will use. On many days the hens should lay from 12 to 15 eggs, perhaps more. There is sure to be a good sale for these eggs at a very satisfactory price, if the poultry keeper cares to deliver them properly packed in egg boxes. Care should be taken to have them absolutely fresh, for a single bad egg is sufficient to ruin the seller's reputation as a reliable hen man. The parcel post offers an excellent way of shipping eggs to a few city customers.

It pays to put eggs of the same color and size in a box, as appearance counts for much. If an egg becomes dirty in the nest or afterwards, it should not be washed, but wiped with a damp cloth. Egg shells are porous. Clean nests are important because an egg is moist when it is laid, so that dirt adheres to it. Many commuters have regular customers in the buildings where they are employed and the square, neatly wrapped parcels so often seen in the hands of incoming suburbanites spell, to the initiated, fresh eggs.

CHAPTER IX

RAISING FANCY POULTRY AS A PASTIME

PROBABLY no hobby or pastime occupies the leisure moments of so many doctors, ministers and other professional men as the raising of thoroughbred poultry. Hundreds of business men, too, and many women breed fancy stock because they enjoy owning and working with a flock of aristocratic birds. A visit to any poultry show will lead one into the company of people from many different walks in life, but all finding a common interest in well-bred fowls.

Breeding high-class stock does not necessarily mean, though, that it is to be exhibited. Many people are satisfied with the pleasure which comes from the owning of good birds. The number of amateurs entering a few birds in the big shows is constantly growing, however, and the judging is followed with keen interest by men of wealth and position, who are in the game only because they enjoy the good-natured rivalry and competition. The fraternity of poultry fanciers covers the whole country and its members are enthusiasts. They are organized into a national body known as the American Poultry Association and most of the popular breeds are represented by specialty clubs.

In order to have poultry eligible for admission to a show, one must work along somewhat different lines from the utility breeder. The birds must conform to certain requirements set down in a book called the American Standard of Perfection and if they possess various defects in form or otherwise will be disqualified at the start and receive no consideration. The birds in some shows are judged by comparison and in some shows scored. Scoring is an advantage to the amateur, as it shows him in what points his birds are weak.

If money counts with the amateur, the ability to breed birds scoring high and winning important prizes will prove a source of no little profit. The writer enjoys the friendly acquaintance of a clergyman who keeps about 100 thoroughbred Plymouth Rocks on a town lot. A few years ago he began showing a few birds each year and was gratified to be awarded a number of premiums. Other breeders noted his birds and his winnings and began writing him for eggs and stock. Now his bank account is annually swelled to a substantial degree — being a minister, it was never very large — by the profits he receives from his fancy fowls.

However, it takes skill and experience to breed prize winning poultry. Some people never acquire the knack, while to others it seems to come naturally. The best way to begin is to buy a trio of carefully bred birds of the breed decided upon from a man who has a well-earned reputation at stake. Fifty dollars is not too much to pay for a male and two hens of really first-class stock. If that is more than the beginner can afford, he can buy less high priced birds, of course. An even less expensive way to begin is to buy a setting of eggs from a pen of high-grade birds. Starting with the best stock one can afford, simply puts one that much farther ahead.

Having secured birds from a good strain, the amateur who works intelligently will seek to perpetuate the qualities of that strain. To carelessly introduce the blood of another strain would be rank folly. If a trio of birds has been purchased from a breeder who is wholly dependable, the amateur may be reasonably sure that the mating will produce good chickens. The pullets hatched may be bred to the original male the next season and again the following Spring as mature hens. Then additional hens may be secured of the breeder from whom the first purchase was made and mated with a cock descended from the original male or with that very bird. This plan will prevent too close inbreeding.

The cockerels used for breeding ought to be practically counterparts of their sire, this being an indication that the points which make the latter desirable are well fixed. For the rest, the amateur breeder must learn by experience and study how to make his matings in order to get the best results. By keeping a copy of the Standard of Perfection, costing $2.00, close at hand, he will be able to learn just the shape, weight, color and markings which a perfect bird of his favorite breed would have to possess. This will be his guide.

It often happens that hens bred strictly for fancy purposes will not lay as well as strains developed for utility. Egg production is neglected in order to secure certain physical characteristics. This is natural and the amateur should not expect to develop a strain along both utility and fancy lines. Moreover, the fancier wants to give his chickens every possible advantage and so will not force his breeding stock for eggs. He will, on the contrary, try to keep his hens in the best possible condition, so that the eggs will have a high percentage of fertility and produce robust youngsters.

Matings should be made soon after the first of January and only a few hens kept with each male. While the utility poultryman may keep his fowls in a single large flock, the fancier will need several pens, so that it will be easy to keep his matings separate. It would be a calamity if occupants of the different pens should become mixed, even for a day.

After the breeder acquires something of a reputation, he finds it an easy matter to sell his eggs for hatching purposes and at a much higher price than they would bring in the market. Hundreds of amateurs with no more than a local reputation have no difficulty in disposing of a considerable number at a dollar for thirteen. Indeed, if a man has a flock of particularly good looking hens of an attractive breed, he usually finds a local demand for hatching eggs, even though he does not pretend to be a fancier. Indeed, one dollar a sitting is commonly paid for eggs from a strain bred solely for high egg production. A small advertisement in one of the papers often helps to bring in a few dollars for eggs and if the amateur starts with a breed not common in his locality, he will be sure to be asked for setting eggs.

The chickens raised must be carefully culled and on this point the amateur will need advice the first season. Not more than fifty per cent. can reasonably be expected to prove of value as fancy stock.

To be a fancier does not necessarily mean to keep what are commonly spoken of as the fancy breeds. Indeed, some of the best-known fanciers specialize in Plymouth Rocks, Rhode Island Reds and Leghorns. The Barred Plymouth Rock breed, by the way, is one of the most difficult in which to secure first-class specimens.

There are many amateurs, however, who are not fanciers, strictly speaking, but who keep the more ornamental breeds because their beauty of plumage, stylish carriage or pert manners appeal to them. The Houdan and Polish fowls, for example, with their curious topknots, have a host of admirers, who try to keep their pens filled with high-grade birds just because they enjoy looking at them and being among them.   The Langshans, Hamburgs, Andalusians, White Faced Black Spanish and Silver Wyandottes are exceedingly ornamental and are prized for that reason.

The Bantams, too, are highly popular among amateurs who breed poultry simply as a recreation. Physicians seem to have a special fondness for these dainty little fowls and some of them are well-known exhibitors.   It is not an easy matter to breed show Bantams, but there is no little fascination in endeavoring to produce prize specimens.

Fowls which are to be exhibited require special training, so that they will submit to the handling of the judge and pose properly for inspection.   This means that they must be worked with from chickenhood and made accustomed to being lifted and carried about.   Patience is required to get them to stand motionless sufficiently long to have their pictures taken while they "look pleasant."   Gently stroking the throat seems to have a soothing effect when the birds are being taught to pose.

White birds which are to be shown need washing before being shipped to the show room.   The usual method is to prepare three tubs of water, one cold, one warm and one lukewarm, a little bluing being added to the cold water. A warm room is needed, for poultry shows come in cold weather; in practice, the kitchen is commonly made use of. The bird is first scrubbed with warm water, using a brush and always rubbing downward. Soap should be used freely, but thoroughly rinsed off in the second tub, the victim being soused in the cooler water. Then a dipping or two in the bluing water will give the finishing touch to the bath. Next in order is a thorough drying with sponge and towel, after which the bird is placed in a coop, the bottom of which is covered thickly with sawdust, and the coop given a location near the fire, but not close enough to cause the feathers to crinkle from the heat.

While fowls with dark plumage are not often washed, all need a certain amount of attention and the man who sends his birds to the show in the best of condition has an advantage over his more slovenly competitor. Whatever the breed, the legs, comb, wattles and lobes should be carefully cleansed with warm water, using soap and a soft sponge.

It will be realized from what has been written, that the breeding of fancy poultry is not an undertaking to be entered upon lightly if one is really ambitious for success in the show room. And yet beginners sometimes have most unexpected success. Time and again a new exhibitor has come before the judges and carried away some of the most important prizes.   And whether one wins or loses, the fascination of breeding thoroughbred poultry does not soon pall.

CHAPTER X

DUCKS, GEESE AND GUINEA FOWL

DUCKS, geese and guinea fowls are not for the amateur who has only a very small lot. The man or woman, however, who has enough land so that the stock need not be confined too closely will find these birds profitable, as well as providing meat for the table at low cost. One variety of duck, the Indian Runner, may even be depended upon for eggs, for it is remarkably prolific, laying from 140 to 200 eggs a year.

It is only of late years that the Indian Runner duck has become popular. Now, birds of this breed are being raised in constantly increasing numbers. Many women are taking up the Runners, seeming to consider that they are somewhat easier to care for than hens. Perhaps this is a fact, for they are hardy and strong, grow quickly and never need to be coddled. It is not at all improbable that some amateurs will substitute them for the more common kinds of poultry in the years to come, for

114

the eggs are large and of good flavor. Some strains lay white and other strains greenish-tinged eggs. Of course the pure white eggs are to be preferred, so that when buying stock, one should be careful to learn the color of the eggs produced. Day-old ducklings may be purchased for about 25 cents each and are easily reared with a hen.

There are three varieties of Indian Runners — penciled, fawn and white and pure white. The penciled Runners represent the English type, but the American standard recognizes, as yet, only the fawn and white variety. Some breeders insist that the penciled birds are more certain to lay white eggs than the American type, and are more prolific, but the latter are more commonly seen. Doubtless all three varieties will be standardized eventually. The whites are rather scarce at present and bring higher prices than the other kinds.

Given proper care, Indian Runner ducks are easy to raise and require no water to swim in, although they demand a surprisingly large amount to drink and dabble in. It is necessary that they have water always before them and in a receptacle deep enough so that they can dip their beaks into the water to the nostril openings, for these openings often become clogged with soft food or mud and the birds are in danger of smothering unless water is at hand. The ducks should be kept in a yard by themselves and in clean, dry quarters free from draughts. If allowed to run with the other poultry they will gobble up more than their share of the food and make nuisances of themselves in other ways. The ducklings must not be allowed out in a shower nor permitted to swim even in a mud puddle until they are feathered out — and they acquire feathers much more slowly than chickens.

Low, rough shelters are sufficient for the ducks. A dry goods box will answer for a small flock and one side may be left open or covered with muslin tacked to a light frame. There should be a liberal supply of sawdust, shavings or straw on the floor for a litter, so that the floor will be dry, and this litter will need frequent renewing, for the webbed feet of the ducks carry much water and mud. The ducks should be confined until the middle of the forenoon in the laying season, for they almost invariably lay their eggs in the morning, often dropping them on the litter wherever they happen to be, but sometimes fashioning temporary nests.

Indian Runner ducks eat about as much as hens.

The ducklings, however, are very greedy, but their rapid growth may be considered sufficient justification for their astonishing appetites. No food should be given for the first 36 hours, although water should be provided and in a dish which the youngsters cannot climb into. The activity of a day-old duckling is surprising to people who are accustomed only to chickens.

Bread soaked in milk or water and sprinkled with coarse sand or chick grit may be fed four times a day for three or four days and then a soft mash gradually substituted. A good mash is made of four parts bran, one part ground oats, one part corn meal, two parts of green stuff and one part of beef scraps. A little chick grit and charcoal may be added. Some breeders put the grit in the water dish, as the ducklings will usually pick it out. The green stuff may be dandelions, lettuce, clover or alfalfa. The mash should be crumbly and not wet. It is well not to include the beef scraps until the ducklings are a week old and to begin with somewhat less than one full part.

When the ducklings are eight weeks old, cracked corn and wheat may be fed at night. Whole corn may be substituted when they grow old enough to eat it easily. When matured, a laying ration made as follows may be fed: Two parts bran, one part ground oats, one part corn meal, one part of beef scraps, one part of alfalfa. Waste vegetables may be added and corn and wheat fed at night, the mash being given in the morning and at noon. Grit and oyster shells should be kept where the birds can have free access to them.

This is the conventional way of feeding ducks and ducklings and serves to keep them in prime condition. Yet simpler methods will answer. Being pressed for time, the writer tried feeding rolled oats dry at frequent intervals and found that the ducklings both relished and thrived on them. He even went so far on many occasions as to sprinkle rolled oats all about the grass run where the ducklings were confined and to leave them from early morning until late in the afternoon, a large covered drinking fountain supplying the water. No bad effects followed, either. Coddling is no more necessary than for chickens. And yet this hit and miss method of feeding is not recommended, if more than a very few birds are being raised.

A yard made of single boards will confine the ducklings at first, and the location should be shifted several times a week. If the ducklings are being cared for by a hen, she may be allowed to jump out and roam around. The youngsters have no use for her, anyway, except as a source of heat, and pay no attention when she gets excited over a bug or worm and tries to call them to the feast. The hen is likely to get disgusted with her charges rather early, but that does not matter, for they become large enough in a few weeks to dispense with her services. Even after the ducks are full grown a low fence will confine them and they give much less trouble than hens. They may be driven like sheep from place to place, as the flock always keep together, and suffer little from vermin.

If given a wide range, the Indian Runners will pick up a large percentage of their rations, for they are excellent foragers. When confined, it is most important that they have green food in abundance. Grass clippings and the refuse from the garden should go into their pen and it is well to grow lettuce, cress or other vegetables for them.

The birds of this breed do not dress as heavy or as attractively as Pekin ducks, which are the table ducks par excellence, but the meat is fine-grained and unsurpassed in flavor. The Runners make excellent broilers. This particular breed has been discussed at length because it is less well known than the Pekin and because it has qualities which commend it especially, the writer believes, to the amateur. Most beginners, at least, do not care to raise poultry purely for the table. The suggestion of taking life is made too emphatic. The Indian Runner may be raised mainly for the eggs produced, with the meat supply as a supplementary item.

It is true that breeders of Pekin ducks often are able to sell the young birds alive, although at a decreased profit. There is no question that Pekins are money-makers; perhaps they are the most famous of market fowl, for they are distinctly a meat breed. Being pure white, they dress to good advantage and the feathers are worth forty cents or more a pound. If properly grown, these ducks are ready for market at ten weeks or a little more, when they weigh five or six pounds.

Pekin ducks and ducklings may be given the rations and general care described for Indian Runners, except that a larger percentage of corn meal is needed when the ducks are being fattened for the table. The ducklings are timid and easily stampeded, sometimes piling up in a corner when startled, with serious results to the birds at the bottom of the pile. Some breeders keep a lantern burning in the house at night, as a partial protection against this sort of thing.

The one other kind of duck which the amateur is likely to keep is the Rouen, which is a particularly good table fowl, but not so popular as the Pekin because of its dark-colored feathers and slower growth. Rouens are hardy, gentle and good layers. They are not so easily stampeded as the Pekins and may be kept in larger flocks. The amateur with a small farm will find a few of them an excellent investment, for they will shift for themselves to a large extent, requiring but little care.

When ducks of any breed are yarded, the question of sanitation becomes an important and sometimes a vexing one. Too large a number should be avoided so that the birds may be shifted from one yard to another occasionally, the yard vacated being spaded or plowed and sowed to a thick-growing crop like rye. Shade must also be provided if there is no natural shelter, and may take the form of strips of burlap or old grain sacks fastened to a light frame. It should not be dense.

Only amateurs living in the country should try

raising geese, for they require more room than ducks and chafe at confinement. Geese are grazing creatures like cattle and need pasture. Meadows and marshes are ideal for them, if they also have access to land which is high and dry. They are very easy to raise, cost almost nothing to keep and bring a satisfactory price when marketed. Many people miss an excellent opportunity to add to their incomes by not keeping a few geese.

Geese live from twenty to fifty years if given the opportunity. Indeed, instances of geese living to be a hundred are not rare. Ganders, however, are likely to become vicious after they reach the age of six years and usually are disposed of when comparatively young, for they have powerful wings and are able to seriously injure women and children. It is not a sign of cowardice to run from an enraged gander.

Often three settings may be secured if the eggs are removed from the nest. Hens may be allowed to hatch the first lots, the eggs laid last being given the goose to incubate. Goose eggs require from 28 to 31 days to hatch, those under geese often hatching quicker than those under hens. As geese do not begin laying until late Winter or early Spring, the goslings usually may be put on grass at once, being given a light mash of bran and corn meal twice a day the first week. If they are to be fattened for the early summer market, the mash should be continued, but otherwise the youngsters will get along very well on grass alone, plus what bugs and insects they are able to secure, although an occasional mash with the addition of cooked vegetables and some beef scraps will help promote growth, as well as teaching the birds to come home from their roaming every night.

Simple sheds to protect them from the biting winds and driving rains are all that geese need. They do not feel the cold. Indeed, a goose will settle down in a blizzard and appear comfortable enough, changing her position only to prevent being buried under the snow.

The Toulouse and Embden geese are the breeds commonly kept. Both are large, massive and attractive. They are much alike in appearance except that the Embdens are pure white, while the Toulouse geese have a large proportion of gray feathers. The Toulouse geese are the more prolific, but the Embdens make the better mothers. Probably the former are to be preferred if yarding is necessary, as they endure confinement with some degree of patience.

Gray African geese are good layers and excellent for the table, as the meat is fine-grained. They can be made to weigh eight pounds in ten weeks, so that they rival the Pekin duck. Many breeders find them profitable. Geese have strong lungs and are prone to exercise them when startled or when strangers approach. Sometimes this is an advantage; every schoolboy knows that a flock of geese once saved Rome.

When it comes to noise, however, the guinea fowl claims attention. Its raucous cry may be heard a long distance and often unpleasantly early in the morning. But then, even the gorgeous peacock offends in this way. It often happens, too, that the cry of the guinea fowl is not to be deplored, for it is highly effective in keeping away hawks and may be depended upon to give the alarm if intruders attempt to enter the poultry house at night. Guineas are being grown in increasing numbers because of the demand created by high-class hotels, clubs and restaurants. Game has become scarce and guinea chickens are the best substitute which has been discovered. Formerly they masqueraded under some other name, but now they are often served frankly as guinea chickens, for their merits have been recognized.

The meat of old guineas is tough, but that of birds a few months old is tender and delicious to a degree few people realize. These birds are well worth growing for the family table, although they may be made decidedly profitable. People who have only a little land can purchase eggs, set them under a hen or two and dress the young birds as they are wanted. By this plan they may be had for the table without the necessity of wintering breeding stock. Mature guineas rebel at being confined, although it is quite possible to keep a few in a comparatively small yard. However, they do but little damage when allowed their freedom, for they do not scratch up the garden like ordinary hens, but walk sedately up and down the rows of vegetables, stopping at frequent intervals to gobble down a bug.

Guinea hens like to make their nests in secret places, and if they are yarded, it is well to provide piles of brush for their use. When at liberty, the location of the nest often may be discovered by watching the male, who stands guard close by during the period of incubation. If left to her own devices, the hen will lay many more eggs than she can cover, so that it is customary to remove them. This proceeding is one which demands caution, for the guinea hen is a wise and suspicious bird. People say she can count to five. Anyway, that number of eggs should always be left in the nest and those which are removed, while the hen seeks provender, must be lifted out with a wooden spoon. The amateur may laugh at this statement, and call it an old wife's tale. He will learn better when he sees nest after nest abandoned.

The eggs removed should be given to hens, although the latter might refuse to take them if they realized the task confronting them. Guinea chicks are strange little creatures. They fairly pop out of their shells when the day of hatching comes, and as soon as they are dried off, they are ready to start out to see the world. Unless boards or netting is placed around the nest, there is danger that the venturesome young guineas will wander away and become lost. Although they grow rapidly and soon become quite capable of caring for themselves, they refuse to be weaned and continue to tag after the hen which has mothered them until they are matured, much to the ill-concealed disgust of biddy.

Guinea fowls, if there are more than two or three, should not be yarded with the other poultry, for they are confirmed mischief makers and will make life a burden for the common hens, chasing them from one end of the yard to the other and driving them away from their meals. As a matter of fact, it is not easy to keep them yarded at all, unless their wings are clipped or the yards covered, for they have well-developed flying powers.

Amateurs usually experience no little difficulty in distinguishing the sexes. The males commonly are larger than the females, possess larger wattles and have some white on their breasts. Then, too, they are not as talkative as their mates and have a shorter note than the "buck-wheat" call of the hen. A pen of guineas usually consists of one male and from six to ten females. The young birds are often sold by the pair and are not dressed. Indeed, the amateur should have no difficulty in disposing of a few birds alive, and so avoid the unpleasant task of killing them. Where there is sufficient land available, it is worth while experimenting with a pen of guinea fowls.

CHAPTER XI

SOME OPEN SECRETS

SUCCESSFUL poultry keeping is not a question of secret methods. If it were, the few who had been initiated into the mysteries of the craft would reap the harvest of eggs, and the rest of us would fare but poorly. There are certain " short cuts," to be sure, which have been widely advertised and some of which have merit. Very few of them, however, are not known in a general way to experienced poultry keepers all over the country.

Take a much-exploited method of picking out the laying hens, which is dependable to a limited extent. The pelvic bones, which must spread to admit the passage of an egg that is being laid, are examined to determine their relative location. If three fingers can be placed between these bones, the hen is supposed to be laying prolifically. If room is found for two fingers, she is laying fairly well. If however,

128

there is barely space for a single finger, few eggs are being produced. My personal faith in this test is not great.

The limitation of the method lies in the fact that it shows only what the hen is doing in the egg-laying line at that particular time. At the season when eggs are coming abundantly, all the hens of the same age and brought up together should be laying and one may feel safe in disposing of such as are derelict. This test may be used when one is reducing his stock in the late Spring, and the hens apparently not laying at that time selected for market. It is a decided saving to get rid of such hens as are not laying in June and July; and of course, it is foolish to keep drones at any season. The condition of the comb and general appearance of a hen are indications to practical poultry keepers as to whether a hen is laying or not.

Amateurs often find it wise to "put down" eggs in the Spring, when they are cheap, for use the following Winter. There are several ways by which eggs may be preserved, but the best one is so simple that there is no reason why it should not be generally used. Silicate of soda, or water glass, may be purchased at any drug store and is diluted with ten times its bulk of water. It is then poured into a crock and the eggs completely submerged. If kept in a cool place, the eggs will remain in good condition for months. It is necessary, of course, that they be fresh at the beginning, and it is better to use eggs from hens with which no male birds are running. These eggs should not be sold, however, for the shells break easily when placed in hot water and the eggs often pop when boiled. And apart from that fact, the wise amateur makes it a point never to sell eggs which he is not sure are less than two weeks old, even to his unsuspecting relatives.

The easiest way to keep track of chicks in order to tell at a glance from which pen they came is to punch the webs of the feet with punches made for this purpose and costing twenty-five cents. Although some breeders wait until the chicks are a month old, it is safer to do the work before the end of a week, for then there is little bleeding and consequently less danger that the chicks will acquire the reprehensible habit of picking each other's feet, as happens when they get a taste of blood. In some instances the feet of chickens have been picked almost to pieces.

The punching should be done quickly and the hole

made well up in the web, but not far enough to injure the bones. By making the holes between different toes and punching both feet it is possible to
make a large number of combinations. It follows,
of course, that a record of these marks must be kept,
or the work will go for nothing. Leg bands are
often used for marking poultry; they are adjustable
to legs of varying sizes and cost but fifteen cents a
dozen or seventy-five cents for a hundred. Each
band is numbered and a record of the numbers must
be kept.

When a broody hen deserts her nest, as a broody
hen sometimes will, the amateur should not become
needlessly alarmed. If the weather is not exceedingly cold or the hen off for several hours before her
defection is discovered, the chances are that the
eggs will hatch, although the chicks may be a day or
two later in coming out. There usually is even
time to go to a neighbor's home to borrow a sitting hen if there is not an extra one at hand. The
new hen's head should be covered when she is being moved to meet such an emergency and the nest
should be made dark when she is put on the
eggs.

If a hen deserts a nest, her action is likely to be

inspired by the presence of lice in greater numbers than she can endure. There is no reason why a hen should not hatch two broods of chicks in succession, if she is given good care. The first chicks may be removed to a brooder or given to another hen with only a few to look after.

Capons may be used to brood young chicks and will care for them with the utmost patience and solicitude. They are better for chicks several weeks old than for very young ones, for they are so heavy and clumsy that they frequently crush their charges unless the latter are active enough to get out of the way, and they are so stupid that they will not think in time to lift their feet in spite of the victim's frantic appeals. Apart from this failing, they are very satisfactory guardians of growing chicks. Some amateurs find it an advantage to have their cockerels caponized, the cost being five to ten cents for each bird. The capons grow very large and make surpassing table fowl, while they may be kept in large numbers in small yards without any sign of quarreling among them. The plan is a good one when it is desired to keep a considerable number of cockerels on hand to be served on the home table from time to time.

When chickens are killed and dressed for home consumption, the ax is usually the weapon relied upon, although professional pickers use a knife with which they pierce the brain through the mouth. The amateur's job, always an unpleasant one, is simplified by using a block into one end of which a spike has been driven. Then a stout bit of cord may be made into a loop passed around the chicken's head and slipped over the spike. Holding the legs of the bird in his left hand, the operator is able to use his ax in his right hand with assurance that the first blow will be the only blow needed, for the chicken will not be able to dodge.

It is as easy to pick a chicken as soon as it has been killed as to let it become cool and then scald it, for the feathers come off quickly while the flesh is warm. The breast and neck should be picked first, as there is most danger of tearing the skin there. A strawberry huller such as is found in most kitchens is highly useful in removing the pin feathers. As soon as the chicken has been freed of its feathers, the carcass should be plunged into cold water and allowed to remain until thoroughly chilled. The experts say that the bird should not be drawn until it is to be made ready for cooking, as it keeps better than when the intestines are removed.

If the amateur prefers to scald his chickens before picking them, the carcass should be immersed in water just below the boiling point, and the water allowed to penetrate the feathers to the skin. As soon as the feathers have been removed, the bird should be thrown into very hot or even boiling water and quickly withdrawn and plunged into ice water, where it may be left several hours. This practice will give the carcass a plump and attractive appearance. Ducks must be left in the hot water considerable longer than chickens in order to have the water penetrate the feathers, but ducks may be picked dry as readily as other poultry. A good plan is to pick the longer feathers of ducks dry and to then scald off the others.

Nest eggs are of value in teaching the pullets to lay in the nests and to prevent several hens piling into one nest. A nest egg is easily made by piercing the end of an ordinary egg and blowing the contents out, afterward filling the egg with plaster paris, warmed so that it can be used readily. When the plaster hardens, a strong, durable nest egg is the result.

When fanciers sell eggs for table use after the breeding season is over, they sometimes plunge them into boiling water, smear them with grease or prick a tiny hole in one end so as to prevent the buyers taking advantage of the low price to set the eggs. All these methods damage the product and the seller's reputation. The only safe way to make sure of the infertility of the eggs sold is to remove the male from the pen in which the hens are confined. Infertile eggs are always best for the table — some people who sell eggs to a high-class clientelle advertise the fact that they market infertile eggs only.

If milk is fed to chickens, it should always be sweet or always sour. It is the alternating of sweet and sour milk which causes trouble. Milk is a splendid food for growing stock and may be used to advantage when clabbered.

In order to get a preponderance of pullets, a cock considerable older than the hens should be used and the breeding pen consist of twenty or more birds. Probably there will be some decrease in fertility, but the object aimed at will be gained in most instances. The amateur hatching his chicks late in the Spring is more likely to get a large number of pullets than the man who hatches extra early chicks.

Although oats are the grain commonly sprouted for green food, barley, wheat or rye may be used just as well. The grain should be covered with warm water and soaked over night. In the morning it may be spread in shallow boxes having drainage holes in the bottom and sprinkled every day with a watering can, using hot water. When the sprouts are an inch or two long they are just right for chickens, but they may be allowed to grow to a length of five or six inches for mature fowls. If the grain is turned or stirred every day, the shoots will not grow as thick as otherwise. This is the much advertised secret of food at fifteen cents a bushel.

If the chicks must be raised in the same runs year after year, it is a great advantage to cover the surface with coal ashes, which should be replaced each season. The ashes keep the soil from becoming foul, tend to keep the ground dry and are relished by the chicks, which consume considerable of them.

When laying hens or pullets must be moved, they should be kept rather hungry for several days and fed in a fresh, deep litter. Being busy hunting food, they will be less excited or disturbed over the change in their surroundings, and the yield of eggs will not fall off materially.

It is an economical practice to feed dried grass clippings in a little rack. An excellent plan is to spread the clippings three inches deep on a section of two-inch poultry wire four feet long and then roll the wire very tightly. The roll should be hung within easy reach of the **fowls.**

CHAPTER XII

INSECT PESTS AND OTHER TROUBLES

IT is to be regretted that the little word lice must be writ so large in a poultry keeper's hand book as is the case. Somebody has said that fleas should be welcomed by a dog because they help him to forget that he is a dog. Perhaps this is true of lice in their relation to the hen; at any rate, they encourage activity on the part of the poultryman, who is obliged for his own comfort to keep his poultry house comparatively free of this pest. Lice spell failure for the lazy amateur, as well as for the professional. The only way to be free of them is to keep everlastingly spraying and dusting — and above all, to keep the premises clean.

Three kinds of lice are found in practically every poultry house, in spite of what some indignant amateurs may say. First, there is the common gray body louse, which feeds at the roots of the feathers and causes the fowls untold irritation. Then there is the head louse, a large and blood-thirsty insect

138

which causes great mortality among young chicks, although it is also found and is propagated on mature birds. Finally, there is the red mite, which preys on the poultry at night and seeks shelter during the day in crevices and corners and under the roosting perch. Turn over these perches in many hen houses, and the under side will be found fairly red with tiny mites. Like the head louse, they suck the blood of the hens and sap their vitality as well as causing extreme discomfort.

The fecundity of hen lice is amazing. Start with one female under favorable conditions and in two months her progeny will number 125,000. Is it any wonder that constant activity on the part of the attendant is necessary?

And yet the amateur need not be discouraged. It is not impossible to keep the pests in subjection. Filth is most favorable to the increase of vermin, and so the house must be kept reasonably clean. Plenty of opportunity to dust themselves will be all the hens ask, as a rule, in order to keep themselves fairly free of lice on their bodies in the day time. The fowls are completely at the mercy of the red mites, however, and the poultry keeper must take a hand in their extermination by making frequent use of kerosene or one of the prepared liquid lice paints. He should thoroughly cover the walls, the perches and the nests once a month the year around and twice a month in Summer. A spray pump will help to minimize the work as well as to economize on material and to reach every part of the house. If detachable fixtures are used, they may be removed in order to facilitate the work as well as to make it more thorough. Special attention should be given the roost and the boards or blocks which support it at the ends. It is best to apply the kerosene or other liquid in the morning on a bright day, so that it will dry and evaporate before the fowls go to roost.

Whitewashing the interior of the house is an excellent plan, especially if a little carbolic acid be added to the whitewash.

If a poultry house is found to be badly infested, it is well to give it a thorough spraying with kerosene or a liquid lice killer and then to dust the hens after they have gone to roost with Persian insect powder (Dalmatian powder) or with a prepared lice powder. If the Persian insect powder is used, care should be taken to have it fresh. A box with a few holes punched in the bottom may be used as a sifter. A ten-cent box with a sifter top may be bought at many drug stores. The hen should be grasped by the legs and held head downward while the powder is dusted into the fluff around the vent and under the wings, favorite haunts of the lice. The powder should be well worked into the fluff and the birds put back on the roosts. It is well to do this work by lantern light, so that the fowls will "stay put." Special attention should be given the roosters, as they are not likely to dust themselves as thoroughly as the hens.

A large proportion of the chicks which perish every season succumb to the ravages of lice. It often is difficult to make the amateur realize this fact, but it is a fact, nevertheless. When the chicks hatch, the head lice at once leave the mother hen for them, speedily exhausting their vitality. The body lice, too, accumulate rapidly. For the latter, powder dusted on the chicks and under the wings of the hen, where the chicks hover, will suffice. This powder will not exterminate the head lice, however; a very little lard or proprietary ointment rubbed on the head of each chick is necessary in order to secure freedom from these pests. Several applications of these insecticides while the chicks are young will be needed.   In fact, there is no likelihood that this work of warring on lice will be overdone at any time.

After lice, the cause of the greatest loss to poultry keepers is tainted ground.   It is not necessary to enumerate the troubles which come from this source.   Indeed, it is not worth while to suggest remedies to use after they come.   The one important point to make is that new ground must be sought at frequent intervals or the old ground kept sweet by plowing, spading, the use of lime or ashes and the growing of green crops.   Air-slacked lime is valuable both outside the house and in.

There are a number of common troubles which may come up to puzzle the amateur.   One of them is frosted combs.   There is no remedy after the comb becomes black, but while the comb is still white with frost it may be held in cold water until the frost has been taken out and then rubbed with carbolated vaseline, drawing the fingers rapidly from the head to the tips of the comb to promote the circulation of blood.

Shelless eggs are the result of too hard forcing for eggs and the use of condiments, among other things. The absence of shell forming material like oyster shells is an obvious cause.   If green food is lacking that may have a tendency to cause this trouble. There seem to be few shelless eggs when a hopper of bran and beef scraps is kept in the pen: Sudden fright may cause the dropping of an egg before the shell has formed.

Egg eating is a bad habit and often difficult to eradicate, for a whole flock may contract it from a single hen.   The original cause usually lies in a lack of shell-making material or of meat.   If a hen with this habit is discovered, she should by all means be removed from the pen.   Sometimes a number of nest eggs scattered about the floor will put a stop to the practice, the hens soon becoming tired of testing their beaks on the hard surfaces.   Another remedy may give better results, although calling for a little more work.   The natural contents of several eggs may be removed and the shells filled with a mixture of soft soap and red pepper, the openings being closed with bits of court plaster.   After breaking a few eggs of this sort and sampling the contents, the hens are likely to be sickened of this habit.   Keeping the nests dark is also advisable.

Feather eating is another bad habit, and may arise either from a craving for more animal food or from lack of exercise. The best remedy is to give the fowls wider range or to use a deep litter in the house and force them to scratch persistently for all the grain they get. Busy hens rarely have time to indulge in these evil practices. Of course, a generous supply of beef scraps should be given, if the trouble may be traced to a lack of sufficient meat in their rations.

Scaly legs are caused by a parasite and the remedy is three tablespoonfuls of lard, two of kerosene oil and one of glycerine, which should be mixed warm and two drops of carbolic acid added. After washing and drying the legs of the birds, this mixture should be applied generously and when warm. Two applications a week for a month will usually prove sufficient.

Sick birds are not worth bothering with, as a rule. The amateur who keeps his flock well housed, well fed and in a sanitary condition will have little trouble with disease, anyway, if cautious about introducing strange birds.

CHAPTER XIII

THE YEAR'S WORK, MONTH BY MONTH

JANUARY

THE hen's working day is short. Be sure that she is kept busy every minute.

Give the evening meal at least an hour before the hens go to roost. Whole corn is the best evening ration at this season. If wet mash is used, feed it early and give a feeding of corn an hour later.

Gather the eggs several times a day; otherwise they may freeze.

If the fowls are closely confined, it is well to shovel some of the snow away from the door so that they can get out on the ground a few hours on bright days.

Don't make the mistake of shutting the house up tightly, even at night. Fresh air in abundance is the one thing needful.

The beginner starting this month can order breed-

145

ing pens, say from six to ten females and one cock
bird.   It pays to buy high-grade stock to breed from.

FEBRUARY

Orders for eggs to hatch should be put in early,
even though delivery is not desired until March or
April.

Incubators should be bought this month, although
the first of March is early enough for the amateur
to start them, unless eggs from heavy breeds like the
Cochins and Langshans are to be set.   It pays to
make a careful study of the incubator question before
making a purchase and it does not pay to buy a cheap
machine.

Eggs to be used for hatching should be gathered
several times a day and kept at a temperature of between forty and sixty.   Eggs over two weeks old
should not be used.

If one is breeding fancy poultry, the first of this
month is none too early to make up breeding pens.

If the dry mash is not being eaten freely, yet is
sweet and inviting, cut down on the supply of whole
or cracked grains.

If there is glass in the house, wash it.   If curtains are used, brush them thoroughly several times a week.

Dampness and draughts are to be watched for. Dry, cold air is much less to be feared.

Keep a deep litter on the floor, but fresh enough so that the grain will disappear from sight. Old litter becomes packed down hard.

MARCH

This is the hatching month for the amateur. Set the eggs for breeds like the Rocks, Wyandottes, Orpingtons and Rhode Islands Reds the first of the month. Several weeks later will answer for the smaller breeds like the Leghorns and Anconas.

When using hens, set several at one time, so that the chicks will come off together, making it possible to double them up and so release some of the hens.

Even when an incubator is used, it is a good plan to set several hens at the same time. Then the eggs under the hens which test fertile may be used to replace those tested out of the machine. This is the way to make the most of a small incubator.

Be sure that the hen is dusted with lice powder several times while she is sitting, the last time just before the chicks hatch.

Spade up the poultry yard as soon as the ground can be worked. If it has been neglected, scrape up and remove an inch or more of the surface dirt, which is sure to be very foul.

The strongest chicks will come from hens which have not been forced hard for winter eggs. Selecting hatching eggs from those laid by overworked pullets is not a wise plan.

Duck eggs usually run fertile this month and this is early enough for the amateur to set them, especially those laid by Indian Runners.

Study the brooder catalogues. Outdoor brooders will be satisfactory for chicks hatched as late as March. Small colony houses with a good hover set in each give good results. After the need for heat has passed, the hover may be removed and the chicks allowed to use the house for a coop.

APRIL

Chicks of the smaller breeds hatched this month will mature in time to begin laying in the Fall.

The brooder must not be overcrowded and requires close watching. The actions of the chicks are the best indications as to temperature.

Sunlight and fresh air are most important if a large percentage of chicks is to be raised. Look to the ventilation. Often muslin can be substituted for glass in at least a part of the window.

The little chicks should get on the ground as soon as possible. A little yard of poultry wire can be used at first, but must be round, so that the chicks working their way along the fence will soon find themselves back in the warm hover.

Chicks with hens require less attention, but biddy should not be allowed to drag them through the wet grass and they must be treated for lice frequently, using powder on their backs and a bit of grease on their heads.

Clean the incubator and give it a thorough sun bath before filling it a second time.

All litter should be cleaned out of the poultry house this month. If there is an earth floor, it is well to remove an inch or more of the surface and replace it with clean sand. This is also a good time to remove all fixtures and give the house a thorough renovation. It may be whitewashed, sprayed or washed with kerosene or treated to a coat of prepared lice paint. Some of the latter are very efficacious, one application sufficing for several months.

Remove the broody hens to the breaking-up coop the first night they are found on the nest, if they are not to be set. The longer they sit, the more difficult it becomes to make them forget that they want to sit.

It is a simple matter to start in the poultry business this month by buying a few broody old hens and several sittings of eggs from good stock.

MAY

Day-old chicks purchased the first of this month will begin laying early in the Fall, if they belong to the smaller breeds. This is a good way for the amateur to begin. The chicks may be put under a broody hen at night, raised in an ordinary heated brooder or grown in a fireless brooder. The latter will give satisfaction now that the weather is fairly warm.

When the breeding season is over, the male birds

should be disposed of, unless of special value. It is poor policy to feed a lot of useless roosters all Summer.

The chicks should be carefully shut in at night, so that they will not fall victims to rats, skunks or thieving cats. It is better to use wire netting with a very fine mesh on the windows and doors, however, than to shut out the air.

If there are hawks about and the chicks are on range, provide numerous brush heaps where the youngsters can find shelter.

Dwarf Essex rape may be planted this month and will make excellent green feed in a few weeks. It should be cared for like cabbage. When the plants are a foot high, break off the leaves, and they will grow again.

If chicks of different ages are running together, the small ones may be crowded away from the feed boxes. To avoid this, make a covered frame of slats sufficiently far apart so that the little chicks can pass through, while the older ones can not. Feed the babies in this frame and they will be able to eat in comfort. The same method may be used to prevent the old hen wasting the rations of the chicks.

JUNE

Give the incubator a thorough cleaning before putting it away. Empty out the oil and remove the old wicks.

The growing chicks must have shade. So, for that matter, must the laying hens. In a Pennsylvania farmer's bulletin, Mr. A. Theo. Wittman advocates planting Jerusalem artichokes in the poultry yard, and the plan seems a good one. They will propagate themselves from year to year and the fowls dislike the taste of the leaves too much to eat them.

Fresh water in abundance is needed for hens and chicks alike. Labor may be economized by using a kerosene or other barrel which will hold several gallons. The barrel should be elevated on blocks and a small hole bored near the bottom. A plug with a groove in one side may be driven into this hole and will allow water to drip slowly into a basin beneath. The amount of water escaping may be regulated by the size of the groove in the plug. Of course, the barrel should stand in a shaded spot.

Remember that cleanliness is exceedingly important in hot weather. All the feeders and drinking dishes should be cleaned daily and twice a week should be taken to the house and scalded.

The chicks should have beef scraps always before them and should be given green food of some kind daily, if confined.

It is well to spade up a portion of the yard each week. A little grain dug into the ground will encourage the birds to take exercise.

Begin getting rid of the old hens this month. Those which obviously are not laying and those which are persistently broody should go first.

JULY

Take extra precautions against lice this month. They multiply with amazing rapidity in hot weather and the hens are not in good condition to resist their attacks.

If the house is very hot, make an opening in the back just under the roof in order to have a cross current of air.

The old hens should be disposed of as fast as they stop laying. Feeding largely of corn will help to put them in good condition for market unless the weather is very hot.

Eggs should be gathered several times a day so that they will not be incubated by the laying hens.

The nesting material should be renewed at frequent intervals and the nest boxes thoroughly cleaned each time.

Treasure your lawn clippings. They are excellent to feed now or to dry and feed next Winter.

Sometimes changing from the regular laying mash to the growing mash fed the chickens will help to increase the yield of eggs in the laying house.

AUGUST

An off month for the poultry. The hens are molting and the production of eggs not large.

Continue to market the old hens as fast as they stop laying. Get rid of surplus cockerels, too.

The supply of corn should be reduced considerably in extremely hot weather.

The molting hens should not be annoyed by a rooster, but should have a shaded yard and cool earth to dust in. Loam is better than dust for the hen's dry bath.

Dwarf Essex rape may still be sown this month to give a supply of late green stuff. Slight frosts do not hurt it.

Be sure that the chickens are not crowded at night.

When confined, pullets and cockerels should not be allowed in the same flock after it is possible to tell them apart, even though hatched late.

SEPTEMBER

An excelient time to build the new poultry house, so that it will be well dried out before Winter comes.

Haul clean sand into the old houses and get them into condition for the season's pullets.

If the chickens must be confined on account of bad weather, give them litter and throw grain into it, which will help to keep them interested. When chicks that have been accustomed to free range are closely confined, they fret themselves fairly thin.

Let the pullets be used to open air houses from the first and they will grow a heavy coat of feathers, so that they will be ready for fresh air laying houses in Winter. In other words, they will be inured to the cold.

This is a good month for a beginner to start with mature birds, which often can be secured at a bargain. If the amateur wants to build up a good strain, he can afford to buy yearlings from a reliable breeder. He may get fewer eggs than from pullets, but he will be prepared to hatch out a fine lot of chicks in the Spring.

Plow or spade the garden and plant rye for the hens to feed on in Winter.

OCTOBER

Leaves make cheap litter. It is a good plan to gather many bags full and to store them in a dry place.

The pullets should be in their permanent quarters by the first of this month, for it is poor policy to move them after they commence laying.

Start feeding rather more heavily and induce the birds to eat dry mash freely.

If the pullets are slow in laying, give them a warm crumbly mash several times a week. If a daily mash of table scraps is given, no stimulant will be needed probably, but a teaspoonful of mustard to the mash for 25 hens may be added. A little salt is a help. Also, a handful of hemp seed may be thrown into the litter once or twice a week.

Keep litter in abundance on the floor and make the pullets scratch for their grain. Exercise and contentment are very important.

Make note of the pullets which lay first, if possible, and put a band on one leg of each. These will be the hens to breed from.

Don't hold the cockerels any longer, unless wanted for your own table.

NOVEMBER

If the pullets are laying freely, you will know that your season's work has been properly done. If they are not; well, what was the trouble? If you have kept a record of your work, that will help you to decide.

More corn may be fed now. Indeed, it may well be fed exclusively for the evening ration.

This is a fine month to sell eggs, but a poor one for dressed poultry. There should be no old fowls left to sell.

Busy hens make a full egg basket. A handful of millet scattered in the litter occasionally will be an extra inducement for the hens to scratch.

Be sure that the pullets have plenty of green food. Cabbages are not the best, for they do not add to the flavor of the egg. However, they will answer, but do not make the fowls jump for them. Mangels are better. Cut alfalfa may always be purchased of grain dealers.

DECEMBER

It is often possible to buy a high-grade breeding male this month at a reasonable price, especially a *yearling*.

The hens which remain on the roost last in the morning and go on first at night are not likely to be good layers. Watch your birds at this season. Go into the house at night and feel the crop of each hen. Note those hens whose crops are only half full and test them several nights in succession. If the same condition is found, you will be safe in saying that those hens are not laying.

Scald the water dishes several times a month, even in mid-Winter. By giving the water warm, you will not have to fill the dishes so often, in freezing weather.

Beware of new corn. You can feed it freely, however, if you first put it in the oven and parch it. The hens will relish it, too, especially if it is fed warm. In fact, warm corn is in the nature of a gentle stimulant, although not of sufficient value to warrant the work of preparing it except in the case of a small flock.

APPENDIX

EXPERIMENT STATIONS

AMATEUR poultry keepers will find it greatly
to their advantage to keep in close touch
with the experiment stations nearest them.
The various stations are located at the places named
below:

Alabama — Auburn, Union-
town, and Tuskegee.
Alaska — Sitka.
Arizona — Tucson.
Arkansas — Fayetteville.
California — Berkeley.
Colorado — Fort Collins.
Connecticut — Storrs and New
Haven.
Delaware — Newark.
Florida — Lake City.
Georgia — Experiment.
Hawaii — Honolulu.
Idaho — Moscow.
Illinois — Urbana.
Indiana — Lafayette.
Iowa — Ames.
Kansas — Manhattan.
Kentucky — Lexington.

Louisiana — Baton Rouge,
New Orleans, and Calhoun.
Maine — Orono.
Maryland — College Park.
Massachusetts — Amherst.
Michigan — Agricultural Col-
lege.
Minnesota — St. Anthony
Park, St. Paul.
Mississippi — Agricultural Col-
lege.
Missouri — Columbia and
Mountain Grove.
Montana — Bozeman.
Nebraska — Lincoln.
Nevada — Reno.
New Hampshire — Durham.
New Jersey — New Bruns-
wick.

New Mexico — Mesilla Park.
New York — Geneva and Ith-
    aca.
North Carolina — Raleigh.
North Dakota — Agricultural
    College.
Ohio — Wooster.
Oklahoma — Stillwater.
Oregon — Corvallis.
Pennsylvania — State College.
Porto Rico — Mayaguez.
Rhode Island — Kingston.

South Carolina—Clemson Col-
    lege.
South Dakota — Brookings.
Tennessee — Knoxville.
Texas — College Station.
Utah — Logan.
Vermont — Burlington.
Virginia — Blacksburg.
Washington — Pullman.
West Virginia — Morgantown.
Wisconsin — Madison.
Wyoming — Laramie.

POULTRY LITERATURE

THE following bulletins are for free distribution and copies will be sent to any address on application to any Senator, Representative, or Delegate in Congress, or to the Secretary of Agriculture, Washington, D. C.:

No. 41. Fowls: Care and Feeding.
No. 51. Standard Varieties of Chickens.
No. 64. Ducks and Geese.
No. 128. Eggs and Their Uses as Food.
No. 141. Poultry Raising on the Farm.
No. 177. Squab Raising.
No. 182. Poultry as Food.
No. 200. Turkeys.
No. 236. Incubators and Incubation.
No. 287. Poultry Management.

The following bulletins contain short articles on the subject mentioned, while those above are entirely devoted to poultry.

Raising Geese for Profit — No. 65.
Feeding Poultry — Nos. 84, 97, 107, 144, 186.
Preserving Eggs — Nos. 103, 273, 296.
Dressing and Packing Poultry — No. 144.
Selling Eggs by Weight — No. 114.
Early Molting of Hens — No. 186.
Cost of Eggs in Winter — No. 190.
Poultry Appliances — Nos. 316, 317.
Fertility of Eggs — No. 251.

Incubation — Nos. 281, 309.
Cause of Death of Young Chicks — No. 309.
Healthy Poultry — No. 305.
Snow for Poultry — No. 309.
Digestibility of Fish and Poultry — No. 276.
Guinea Fowl — No. 262.

The bulletins of the Bureau of Animal Industry, United States Department of Agriculture, treating on poultry topics, may be purchased of the Superintendent of Documents, Union Building, Washington, D. C.   A circular giving titles, prices, etc., may be had upon application to the above address.   Most of the state experiment stations issue free bulletins, some of which are of great value.

OFFICIAL SCORE CARD OF THE AMERICAN POULTRY ASSOCIATION

Date ....   Variety ....

Owner ....   Sex ....

Address ....   Band No....................

Entry No..................   Weight ....

	Shape	Color	Remarks
Symmetry	....	....	....
Weight or Size........	....	....	....
Condition	....	....	....
Head and Beak........	....	....	....
Eyes	....	....	....
Comb	....	....	....
Wattles and Ear-Lobes	....	....	....
Neck	....	....	....
Wings	....	....	....
Back	....	....	....
Tail	....	....	....
Breast	....	....	....
Body and Fluff........	....	....	....
Legs and Toes........	....	....	....
[1] Hardness of Feather..	....	....	....
[2] Crest and Beard......	....	....	....
Total Cuts..............		Score....................	

[1] Applies to Games and Game Bantams.

[2] Applies to Crested Breeds.

Name of Judge..

Secretary ....

163

SIMPLE TOE-MARKING SYSTEM

I T is a simple matter to keep track of the chicks if they are toe-marked. Then it becomes easy to identify the birds of different hatches and to tell the year and even the month in which they made their appearance. The Bureau of Animal Industry at Washington suggests the following system of markings by which sixteen combinations are made possible. The average amateur will need to use only a few of them, but the diagram indicates the possibilities of the plan:

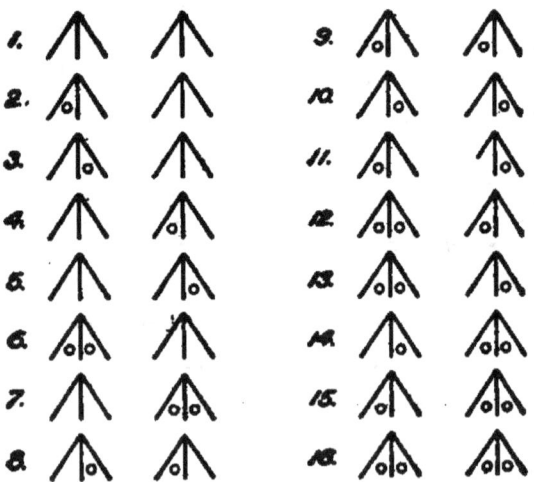

INCUBATION PERIODS

THERE is considerable variation in the time required for the incubation of eggs laid by the different fowls, as the following table shows:

Common hen ....21 days.
Partridge ....24 days.
Pheasant ....25 days.
Guinea hen ....25 days.
Common duck ....28 days.
Pea hen ....28 days.
Turkey ....28 days.
Goose ....30 days.

LIBRARY
COLLEGE OF AGRICULTURE
UNIVERSITY OF WISCONSIN
MADISON

INDEX

167

Corn, feeding new, 157, 158
Curtains clogged with dust, 33
  in fresh-air houses, 27

D

Day-old chicks, beginning with, 5, 150
  cost of, 5
Disinfecting poultry houses, 95
Dropping boards, 41
Ducks, how to dress, 134
  how to feed, 117
  Indian Runner, 114
  Pekin, 120
  Rouen, 121
  shade for, 121
Dust bath, how to make a, 89

E

Eggs, brown, 12, 22
  care of, 104
  cooling, 68
  effect of food on color of, 88
  effect of food on flavor of, 87, 88
  for hatching, 4, 59
  in winter, 84
  preserving, 129
  sale of, 104
  sprinkling, 69
  storing, 88
  testing, 69
  turning, 67
  white, 12, 22
Expenses for stock, 4
  for buildings, 9, 37
Experiment stations, locations of, 159

F

Fancy poultry as a source of profit, 109

Fancy poultry — *Continued.*
  fitting for shows, 111
  kept for recreation, 105
Feeding balanced rations, 46
  beef scraps, 53, 75, 117
  chickens, 74, 81
  ducks and ducklings, 117
  goslings, 122
  grain, 46, 51
  laying hens, 46
  with patent feeders, 97
  milk to chicks, 135
  sprouted grains to chicks, 136
Feeds for chicks, 75
  green, 56, 76, 86, 101, 119
  influence of, on color in eggs, 88
  influence of, on flavor in eggs, 87–88
Fencing, importance of, 102
Foundations for poultry houses, 39
Floors, different kinds of, 38
Frosted combs, remedy for, 142

G

Geese, age of, 122
  dispositions of, 122
  Embden, 123
  feeding, 123
  Gray African, 124
  Toulouse, 123
Grass clippings, how to feed, 137
Grain, kinds to feed, 46
  sprouted, how to prepare, 136
Green cut bone, value of, 89
Green feeds
  alfalfa, 56
  beets, 57
  cabbage, 57
  clover, 56
  for ducklings, 119

GLOSSARY OF COMMON POULTRY TERMS

Brassy — White plumage with yellowish tinge.

Breed — Groups of fowls with distinct characteristics.

Brood — Number of young chicks being raised together.

Broody — Showing a desire to incubate.

Carriage — Characteristic attitude of a bird when at rest and in motion.

Class — Divisions in which various breeds are associated. There are fourteen in the American Standard of Perfection.

Cock — Male bird at least a year old.

Cockerel — Male bird under a year.

Condition — The state of a bird's health, plumage and cleanliness.

Crest — Tuft of feathers on the head.

Crop — Receptacle which receives and softens the food before it passes into a bird's crop.

Cushion — Feathers at the rear of the back of a fowl.

Disqualification — A defect which renders a bird unworthy to compete for a prize. Disqualifications are named in the Standard of Perfection.

Down — The fine hairy covering of chicks.

Drake — The male of the duck family.

Duckling — Young of the duck family.

Ear-lobes — The skin just below the ears. It varies in color with the breed.

Flights — Long feathers of the wings used in flying, but concealed when the bird is at rest.

Fluff — Soft downy feathers on a fowl's posterior.

Gills — Another name for wattles.

Gosling — The young of geese.

Hackle — Long neck plumage.

Hen — Female at least a year old.

Lacing — Feathers edged with a color different from the main color of the wing.

Lopped comb — A comb falling to one side as often seen on Leghorns.

Mandibles — The upper and the lower parts of the beak.

Mottled — Feathers marked with surface spots of another color or shade.

Pea comb — A triple comb having short serrations.

Pen — One male and four females placed together for breeding.

Penciled — Feathers with narrow or concentric stripes.

Poult — A young turkey before the sex can be determined.

Pullet — A female under one year.

Recognized — Conceded as a standard breed.

Rose comb — A solid, low, thick comb, covered with small points.

Rooster — Common term for a male bird, but not used in the fancy.

Saddle — The rear part of the back of a male.

Shank — The part of the leg just above the foot.

Sickle feathers — The long curled feathers at the top of a male bird's tail.

Spur — Pointed or knob-like growth on the inner part of the shank.

Squirrel tail — So called when any part leans toward the neck beyond an imaginary line perpendicular to the back of its junction with the tail.

Under color — The color of the plumage close to the body and hidden by the feathers.

Wattles — Fleshy growths hanging near the beak.

Wry tail — Term applied when the tail is permanently one-sided.

An example of the Tolman fresh-air house. The front is left open the year around, even in cold climates

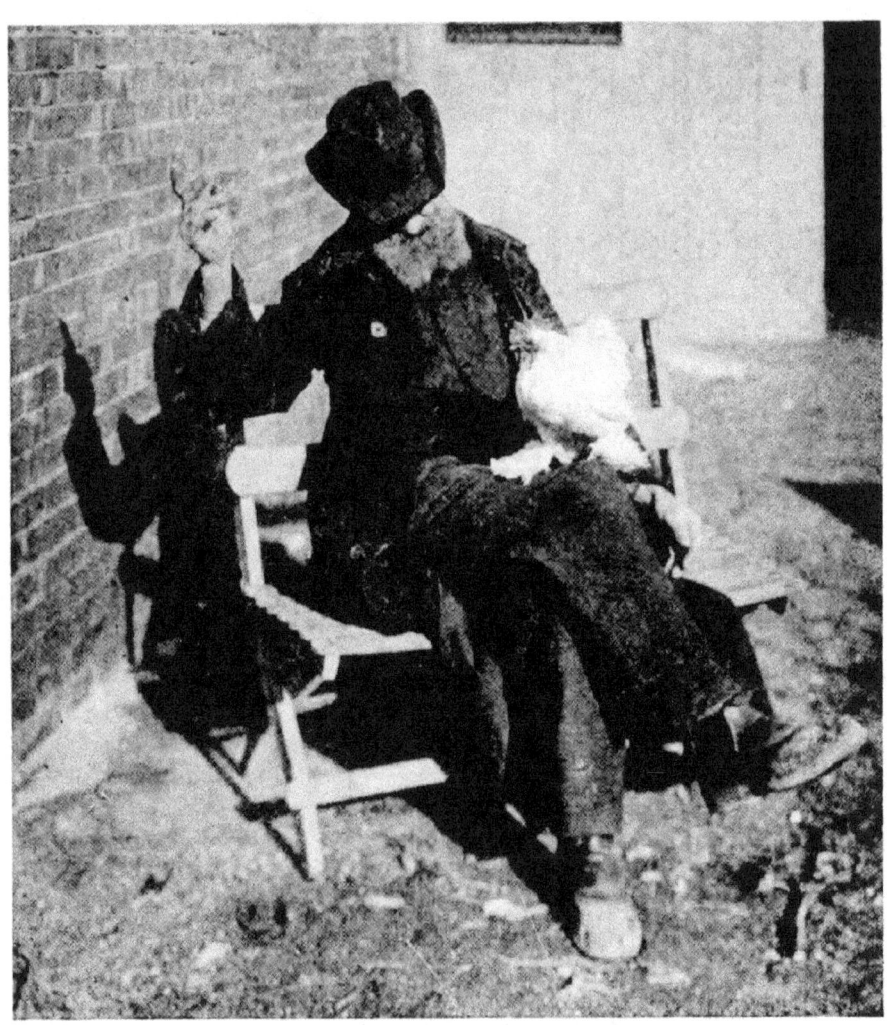
Teaching a White Cochin Bantam to pose. Bantams are
very tame and friendly

On this home plant the birds are confined to the houses winter and summer alike. Pullets are purchased in the fall and sold as market poultry the following season

A flock of Bantams in a low scratching-shed. Very small houses suffice for Bantams, but they need to be dry and without drafts

Indian Runner ducks are remarkably prolific and are being widely bred. It is important to get a white egg strain, as some lay green eggs

Pekins are the best ducks for meat. Curled tail feathers (note the bird at the left) indicate the drakes

Three pearl-gray Guineas and a white one. Young Guineas make the best substitute for game birds

Toulouse geese are noisy, but are profitable if they have
grass land to run on

All light-colored birds need to be thoroughly washed before they are entered in a show

Shade, which the poultry need in summer, is easily provided by growing climbing vines over the yards

A roosting closet, with muslin-covered frames to drop at night in cold weather

The openings, with shutters, at the back provide ventilation on hot nights

The front openings in this semi-monitor type of house have hinged curtains, two to a pen. The house is built of matched boards and has a cement foundation

An excellent colony house for growing chicks. The shutter keeps out rain, while the wired gate lets in air but excludes cats and vermin

Outdoor brooders are practical for the amateur who does not hatch his chickens early. They are hard to manage in very cold weather

A simple fireless brooder in which a few chicks may be
raised in the house

The slanting board prevents the hens scratching the grain
out of the trough

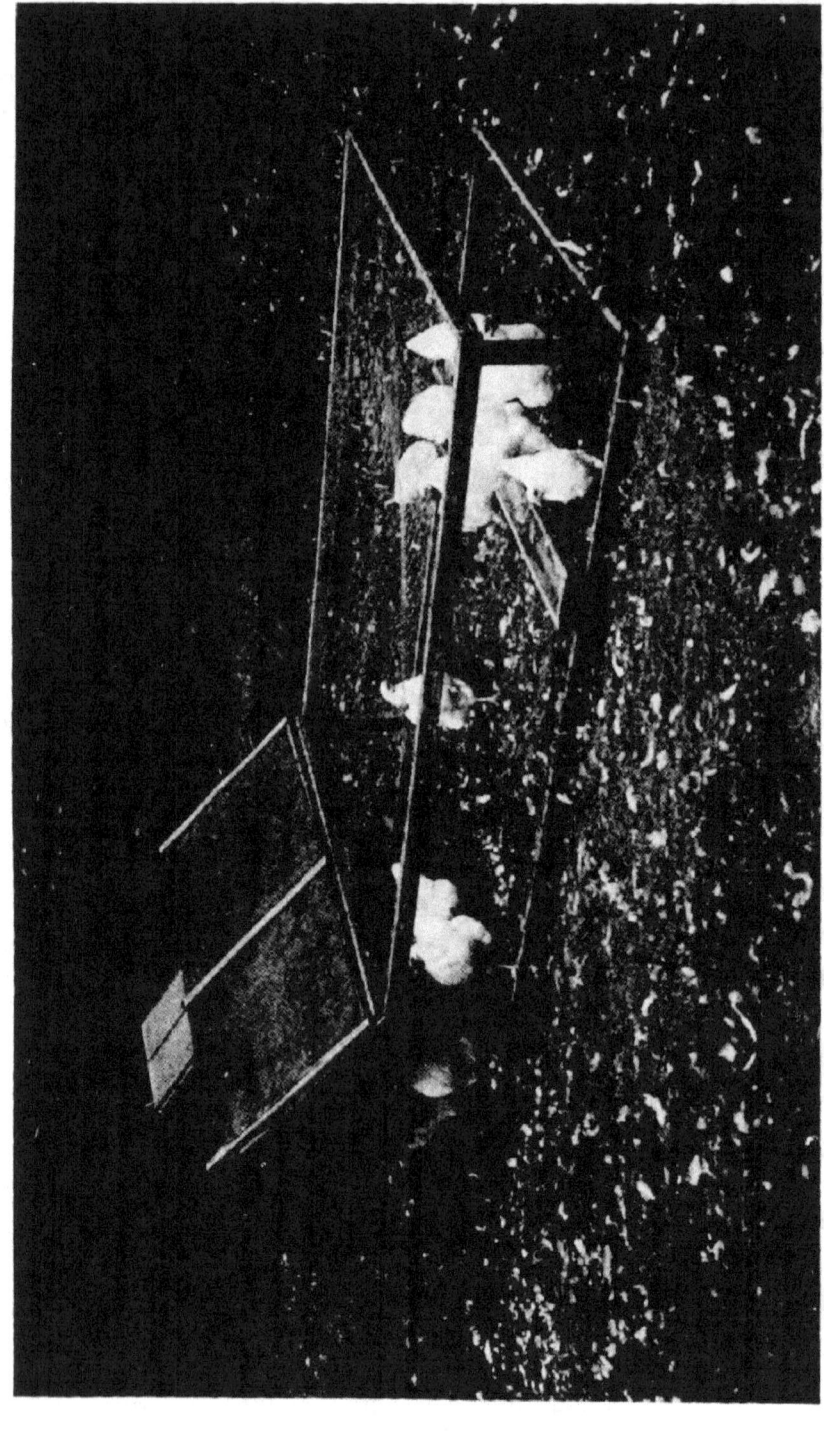

Shelter with yard for growing chicks. It should be moved its width twice a week to keep the birds on fresh ground

Both pleasure and recreation are to be found in caring for a flock of well-bred hens

Incubator houses built of terra cotta hollow tile keep a very even temperature. The idea is a new one

This incubator cellar is too large for the amateur, but illustrates a common method of construction on practical plants

A trap nest and a feed hopper like these may easily be made at home. The illustrations tell the story plainly enough

A simple rack for sprouting oats, to provide green food
in the winter months

Combined exerciser and feeder. When the birds peck at
the bar, a shower of grain is released

"A" houses of the open-front type are inexpensive and serviceable. Usually they are covered all over with roofing paper

An excellent house of the half-monitor type designed by Prof. A. L. Clark of the New Jersey State Experiment Station. Many practical poultrymen are adopting this design

An ideal home poultry plant designed at the Perdue University Experiment Station. The glass window extends almost to the floor and there are muslin curtains for the openings

Shingled house for a small flock. The door is left open
on all fair days

This house is made of building paper laid on poultry
netting, without boarding

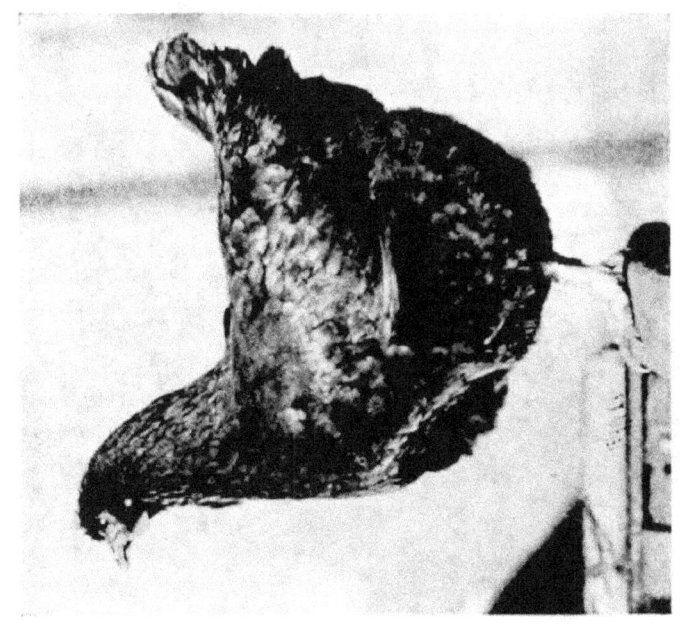

Black Orpington hen. The Orpington
is a very popular breed

Eggs for private customers should be
uniform in size and color

Houdans are of French origin and have large crests. They
are hardy, attractive and easy to keep

White Indian Runner Ducks are handsome and prolific
and seem likely to become as popular as the older varieties

A Silver Wyandotte hen. There are eight
varieties of Wyandottes

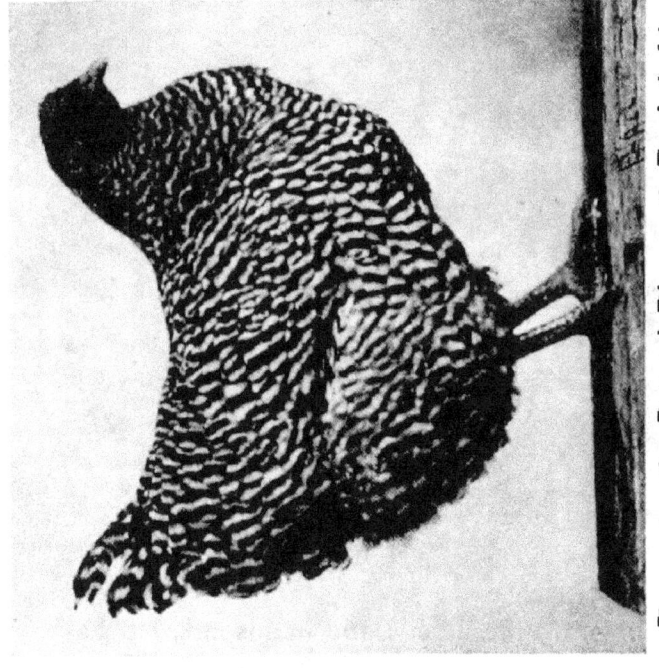

For years the Barred Plymouth Rock held
sway as the great American breed

White Leghorns are the most popular of the strictly egg
breeds. They lay white eggs

Black Langshans are large, docile and handsome, but have
pinkish-white flesh and feathered legs

Among the Wyandottes, the white variety
is the best known and is widely bred

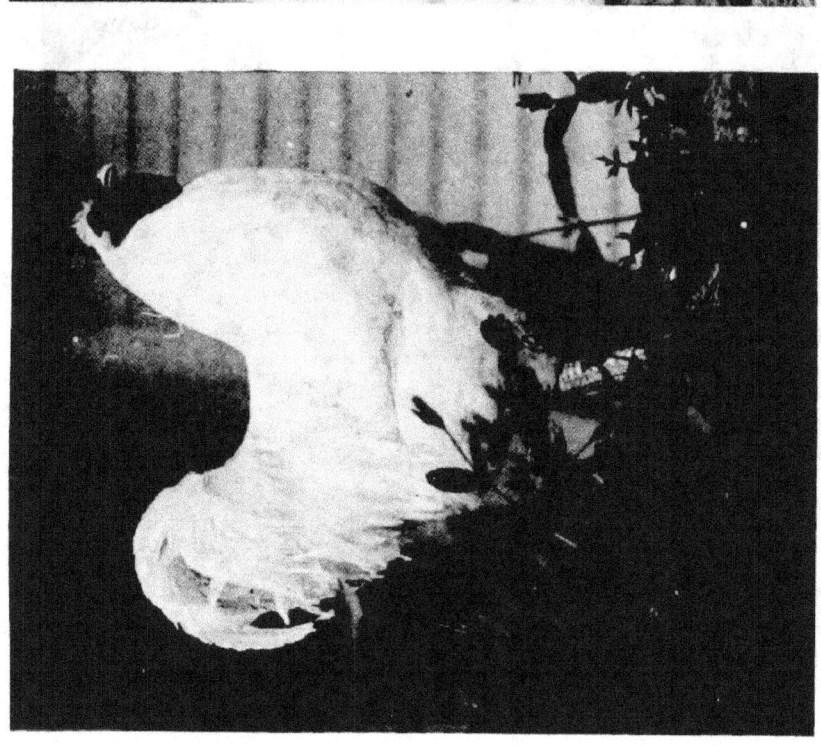

Amateurs find White Plymouth Rocks
very satisfactory. Stylish and handsome

Chicks need fresh air in abundance and should be allowed to get on the ground just as early as possible. It is almost as disastrous to coddle as to neglect them

www.ingramcontent.com/pod-product-compliance
Lightning Source LLC
Chambersburg PA
CBHW081720220526
45468CB00008B/1918